IMAGES
of America

DETROIT FIRE DEPARTMENT

On the Cover: About 100 men and women, many physically handicapped, escaped a fire in this six-story building operated by Goodwill Industries, located in downtown Detroit at 1453 Brush Street between Gratiot and Madison Avenues. This five-alarm fire took place on March 10, 1938. More than 225 firemen fought this blaze with 40 hose lines, six aerial trucks, three high-pressure wagons, and the water tower. This building once housed the McGregor Institute. (Courtesy of Matt Lee.)

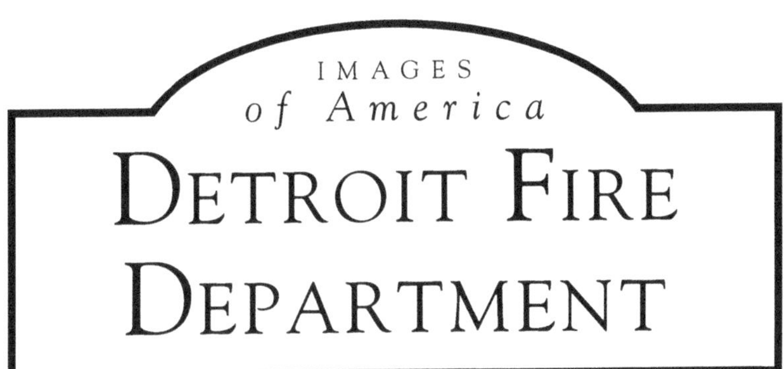

David Traiforos and Arn Nowicki

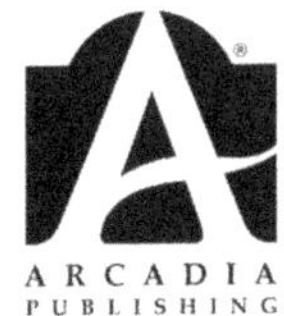

ISBN 978-1-4671-1522-3

Published by Arcadia Publishing
Charleston, South Carolina

Library of Congress Control Number: 2015944653

For all general information, please contact Arcadia Publishing:
Telephone 843-853-2070
Fax 843-853-0044
E-mail sales@arcadiapublishing.com
For customer service and orders:
Toll-Free 1-888-313-2665

Visit us on the Internet at www.arcadiapublishing.com

To My wife, Gloria, thank you for allowing me the freedom to follow my passion. When I am with you, I know that I am home. It's good to be home.

To my children, Andrew, Erik Martin, and Emily, I won't say that I would die for you—of course, I would. After all, I am a firefighter. Rather, let me say, my love for you is so deep, so profound, that I am a better man because of you. Thank you for that.
Love, Dad

Contents

Acknowledgments 6

Introduction 7

1. The Early Times 9
2. 1900s–1920s 19
3. 1930s–1940s 31
4. 1950s–1960s 49
5. The 1967 Detroit Riots 75
6. 1970s–2000 89
7. Detroit Today 107

Roll of Honor 126

ACKNOWLEDGMENTS

This book covers over 150 years of photographed and documented history. It would not be possible without the help of several people who wanted to help document the history of this great organization; they are now our friends. We wish to thank the following people with personal words of thanks: Detroit Fire Department Executive Fire Commissioner Edsel Jenkins, thank you for your permission to allow us to work on this project; fire historian Matt Lee, we thank you for all your knowledge and expertise; the Detroit Historical Society's Adam Lovell, thank you for all your help and letting us go through hours of photographs and historical items; Detroit Fire Department battalion chief Joyce Stoll, thanks for all your help and photographs; Detroit firefighter Jon Soave, thank you for sharing your collection of images; the family of Detroit retired captain Joseph Mancinelli, thank you for sharing all of your father's photographs; Bill Eisner, photographer, thank you for sharing your images; Bill Grimshaw, photographer, thank you for your help and images; Dan Jasina, of *Box 42*, thank you for sharing your photographs; Edward Abair, thanks for the fire badge artwork; Debbie Lyon, we are grateful for all your knowledge of the English language; Karen Schimeck and Bill Day, thank you for use of the weeping dalmatian art work; and Sheryl Jason, of dfdlegacy.com, thanks for your help.

To all the men and women of the Detroit Fire Department (DFD), we thank you for making this the greatest job in the world! And to the readers of this book, thank you for your love and support of the Detroit Fire Department.

INTRODUCTION

In 1805, the village of Detroit was already 100 years old and about to become a city. At that point, there was no organized fire department, but fire regulations had been put in place. A major fire broke out, and as it spread, the community rallied to control the flames with the classic bucket brigade. Exhaustion set in, and the effort was abandoned. Detroit burned to the ground.

Detroiters back then, as today, refused to give up, and the town was rebuilt larger than it was prior to the fire. Indeed, the following year, in 1806, Detroit was incorporated as a city. Its motto and seal recall the devastating fire of 1805. "Speramus Melora" and "Resurget Cinerbus" are Latin for "We hope for better things" and "It has risen from the ashes." The seal depicts two women; one represents the tragedy of the fire while the second the rebirth. Detroit firefighters, upon completion of their training, are presented a chest badge bearing the Detroit seal. It serves a reminder of perseverance under adversity.

The growth of Detroit continued, and for a number of decades it was the fourth-largest city in the United States. The borders spread out to 139 square miles with a population of close to two million residents. After the great fire that leveled the village, a great city was built. To protect the new city, a volunteer fire department was organized. The need for firefighters led to a partial paid/volunteer department. In 1860, the department became professional.

The automobile industry fueled the creation of millions of jobs. And, thanks to labor unions, Detroiters helped create America's "middle class." Below the surface, factors began to chip away at Detroit. As machines started to replace workers, once plentiful jobs became harder to find. Other cities eyed Detroit's jobs with envy and actively wooed companies away while foreign auto companies began importing to the consumers served by Detroit.

Beginning as a trickle, jobs began to leave the city in the late 1950s. The 1960s brought changes to the country. Uncertainty was the new normal for Detroit as summer arrived in 1967. What happened and why can be debated until eternity, but on July 23, Detroit erupted. For five days, the city destroyed itself from within. During that period, Detroit firefighters, assisted by other departments, including one from Windsor, Ontario, fought over 1,682 fires. The loss was staggering. Two Detroit firefighters lost their lives.

In the years following the riots, Detroiters again worked to rebuild their city. The fire department began to reflect its city as race and sex barriers were broken. The emergency medical service (EMS) division was added to give citizens quick and specialized medical care in a pre-hospital setting. Detroit was knocked down but not out.

As the auto companies rebuilt and retooled factories, it did not happen in Detroit. Motown, the city that put the world in cars, was to be reduced to a supporting character. As the jobs left, so did the population. Left behind were the discarded amenities of a larger city. There were vacant factories, stores, schools, churches, and homes. Detroiters worked tirelessly to turn the tide back to prosperity.

On October 30, 1984, the fires began. In the Midwest, the night before Halloween is called "Devil's Night." Prior to 1984, Devil's Night was considered just a night of pranks by children too old for trick-or-treating. In Detroit, it became a celebration of arson, and it had the busiest fire department in the country. The runs taxed the equipment, and Detroit Fire Department (DFD) became known for doing more with less. Today, the city has emerged from bankruptcy, and angels have taken the devil out of Devil's Night.

This book gives a glimpse of how firefighting in Detroit has evolved and added to the area's rich history. The department has gone from horse-drawn steam pumpers to powerful, modern-day firefighting equipment. For generations, men and women have worked in extraordinary circumstances for a common goal: to save lives and protect property.

A true story: a Detroit firefighter arrives for work early one day. Ten minutes into his shift, adrenalin is pumping through his body as he crawls through thick, acrid smoke methodically searching for a victim trapped in a fire. He locates the person and gets them to safety. Just 40 minutes later, he is back at the fire station scrubbing a toilet.

The job is a roller coaster of highs and lows. Photographs will never give the full experience of being a part of the Detroit Fire Department. They cannot capture feeling both exhilaration and terror simultaneously or the heavy burden that exists when every decision can mean the difference between life and death.

We hope this collection will give an overview of the department and help you begin to understand why when Detroit firefighters part ways, they do not say, "Goodbye," but rather "Be Safe."

One

The Early Times

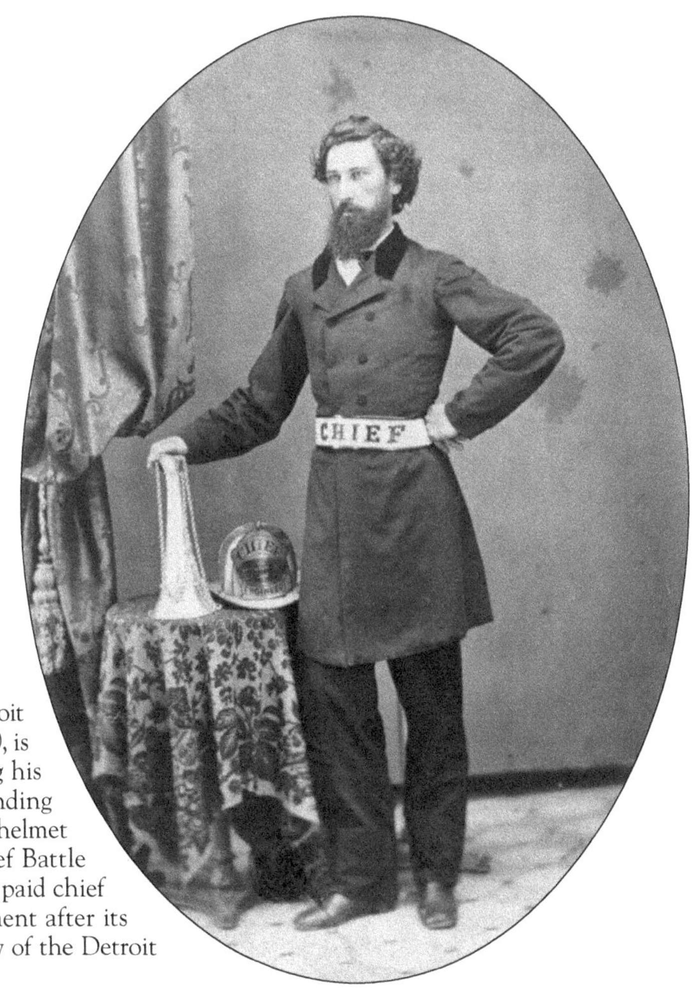

Chief James Battle, of the Detroit Fire Department back in 1870, is posing for a portrait wearing his "Chief" parade belt and standing next to his "Chief Engineer" helmet and speaking trumpet. Chief Battle was noted for being the first paid chief of the Detroit Fire Department after its formation in 1865. (Courtesy of the Detroit Historical Society.)

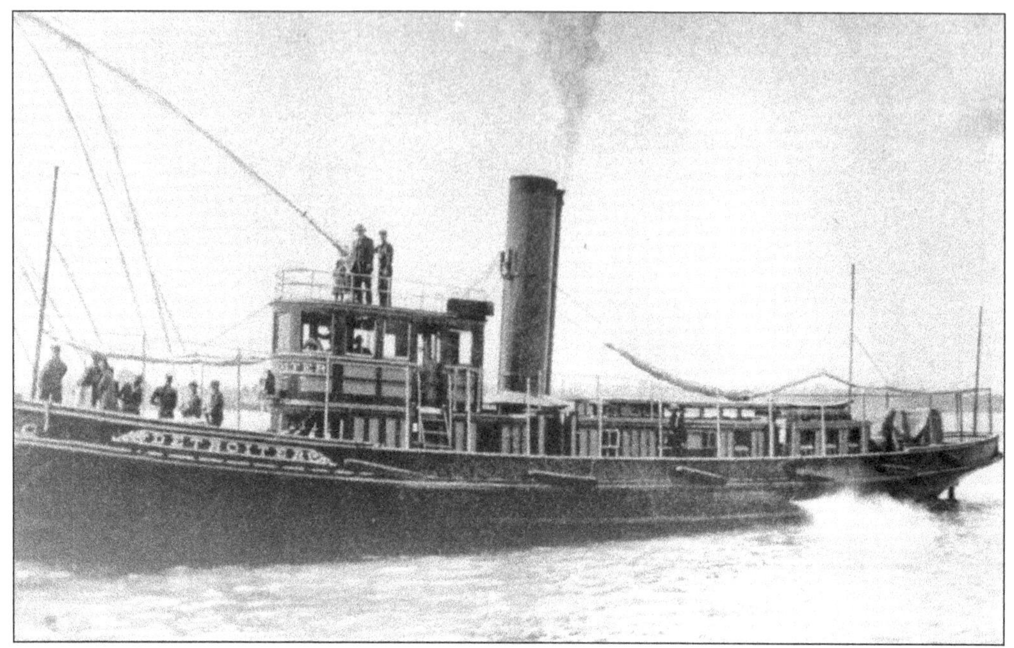

This view shows the port side of the fire tug *Detroiter*. The vessel is underway on the Detroit River. Several crew members are standing on the bow, and one fireman is operating a deck gun above the pilothouse. Detroit's first fire tug was made of wood, built in Toledo, Ohio, around 1893. (Courtesy of the Detroit Historical Society.)

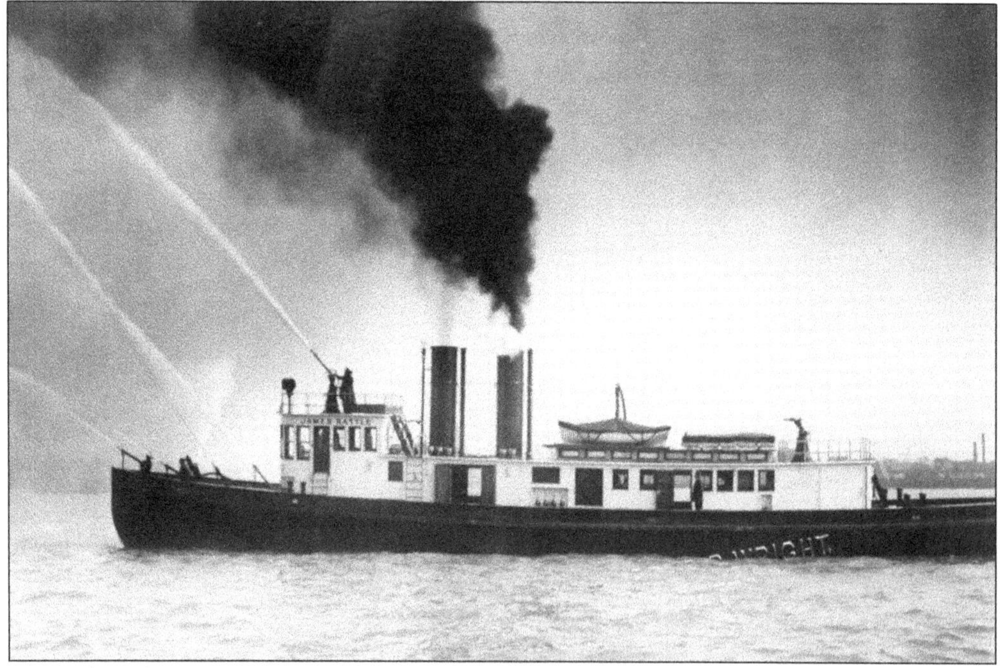

Pictured on the Detroit River, the fireboat *James Battle* was named after the first fire chief of the Detroit Fire Department. Several Detroit firemen are operating hose lines mounted on the bow of the boat, and a fireman is operating a deck gun above the pilothouse. (Courtesy of the Detroit Historical Society.)

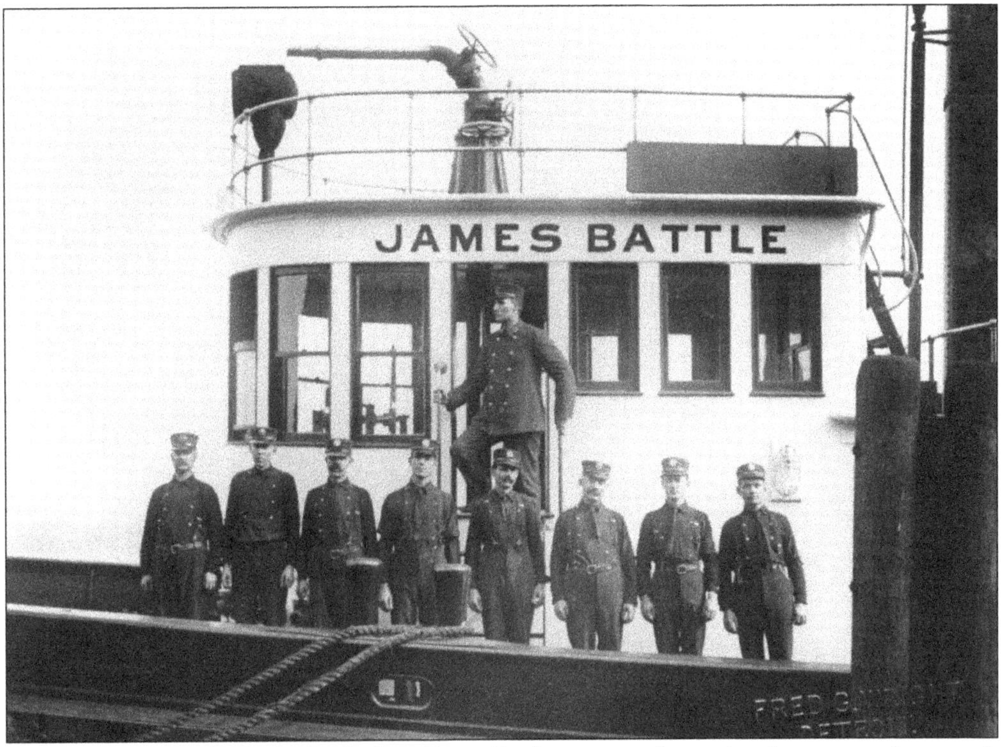

In port and tied to the dock, the fireboat *James Battle* serves as the setting for a group photograph of the boat's crew. The captain of the crew stands in the doorway of the pilothouse while firemen look on. (Courtesy of the Detroit Historical Society.)

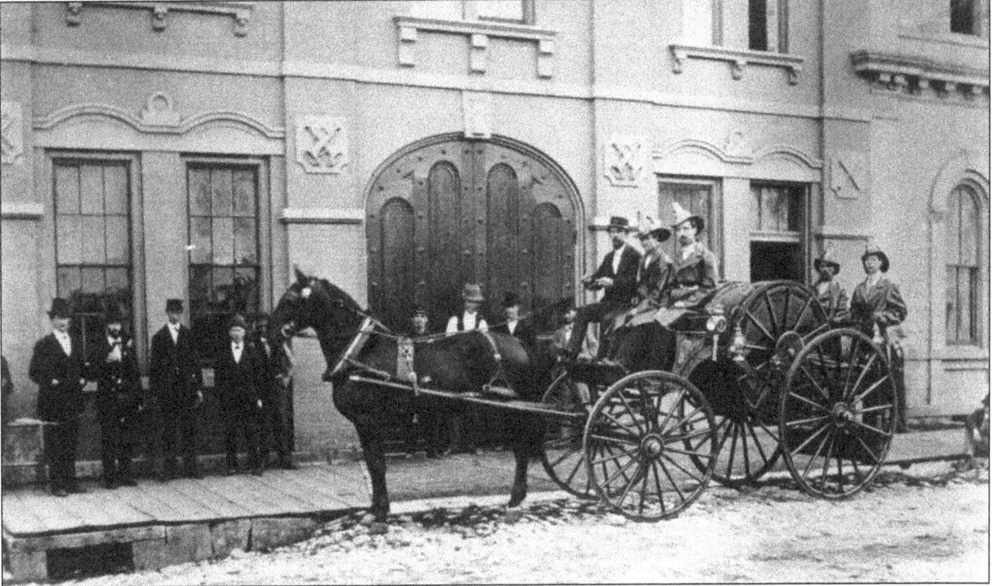

A c. 1860 horse-drawn hose reel apparatus carrying firemen is seen in the street in front of West Lafayette Boulevard. Home to Steam Engine No. 1, the firehouse was decorated with stone blocks with chiseled images of crossed ladders and speaking trumpets. The No. 1 is visible on the lanterns of the apparatus. (Courtesy of the Detroit Historical Society.)

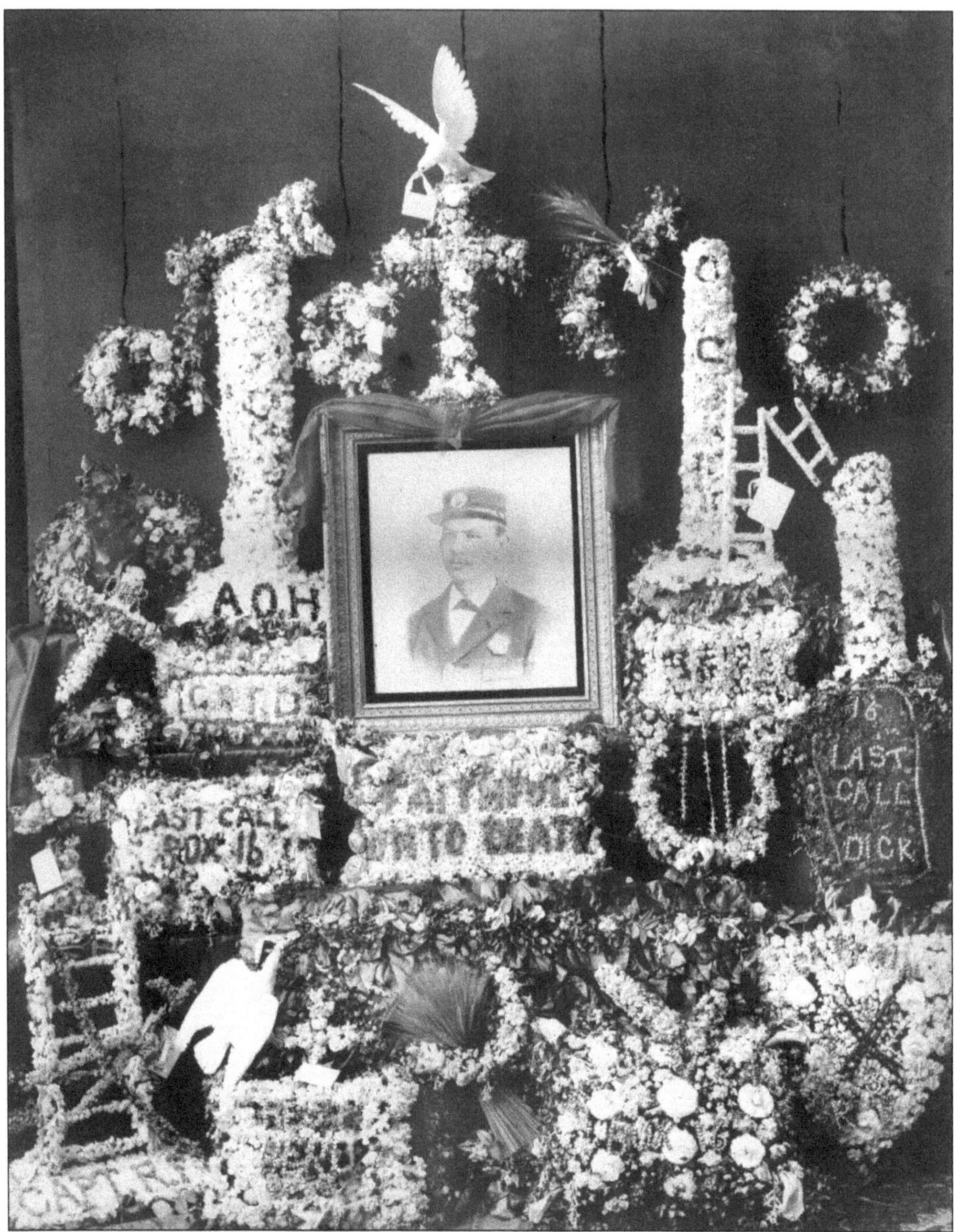

Here is a memorial display for Capt. Richard Filban of the Detroit Fire Department's Hook and Ladder Company No. 3. Filban died in the line of duty at the D.M. Ferry and Company Seed Warehouse fire on January 1, 1886, when a wall fell on him. A framed head-and-shoulders photograph of Filban in uniform is surrounded by a floral display. Wreaths, crosses, doves, ladders, axes, and hooks are among the motifs represented in the display. The following messages are spelled out in flowers: "Faithful unto death," "A.O.H.," "G.R.F.D.," "Capt. R.F.," "Last Call Box 16," and "Last Call Dick." (Courtesy of the Detroit Historical Society.)

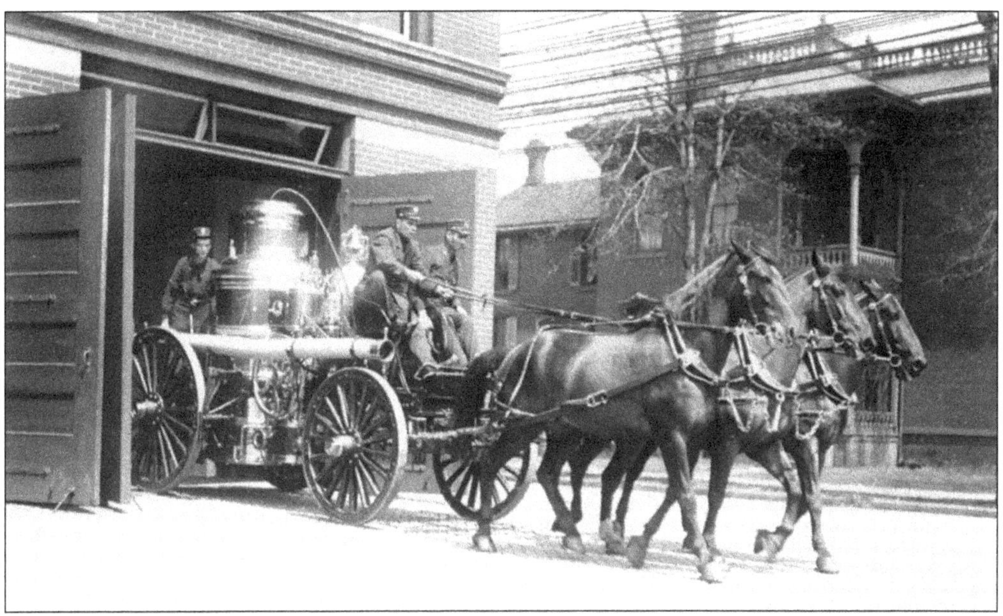

Engine Company No. 9 is seen exiting a fire station with a horse-drawn Amoskeag steam engine. One fireman drives a team of three horses pulling the equipment, a second fireman is seated beside him, and a third fireman is standing on the rear of the engine behind the boiler. Engine Company No. 9 was located at the northwest corner of East Larned and Riopelle Streets. The firehouse was razed in 1961 for urban renewal. (Courtesy of the Detroit Historical Society.)

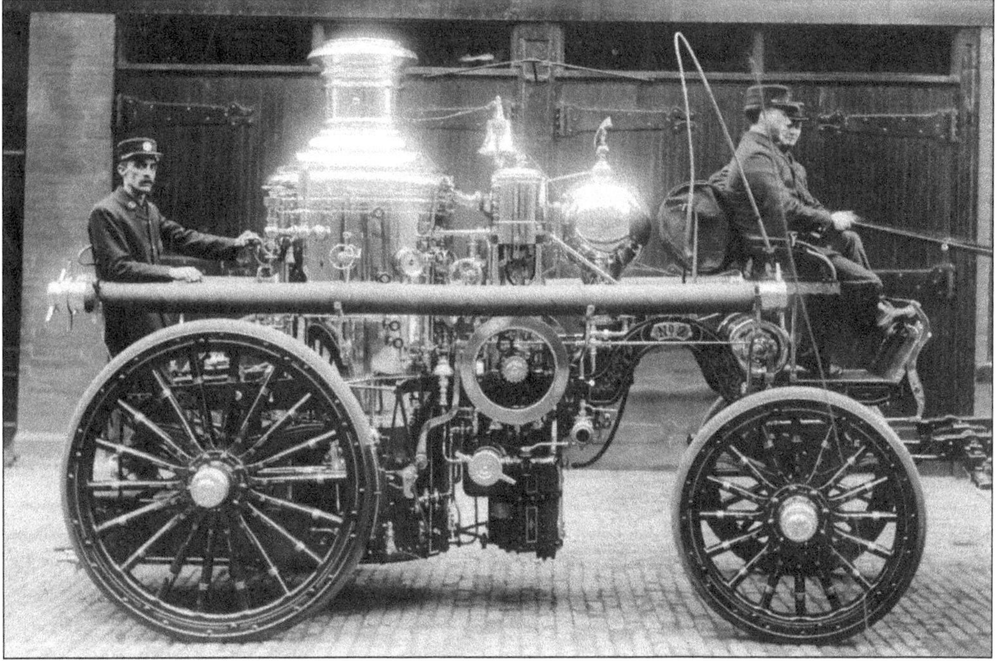

Firemen of Engine Company No. 2 located at St. Antoine and East Larned Streets are seated on their horse-drawn fire engine steamer. They are parked in front of the large bay doors of the fire station. Steam Engine No. 2 was placed in service January 7, 1861. (Courtesy of the Detroit Historical Society.)

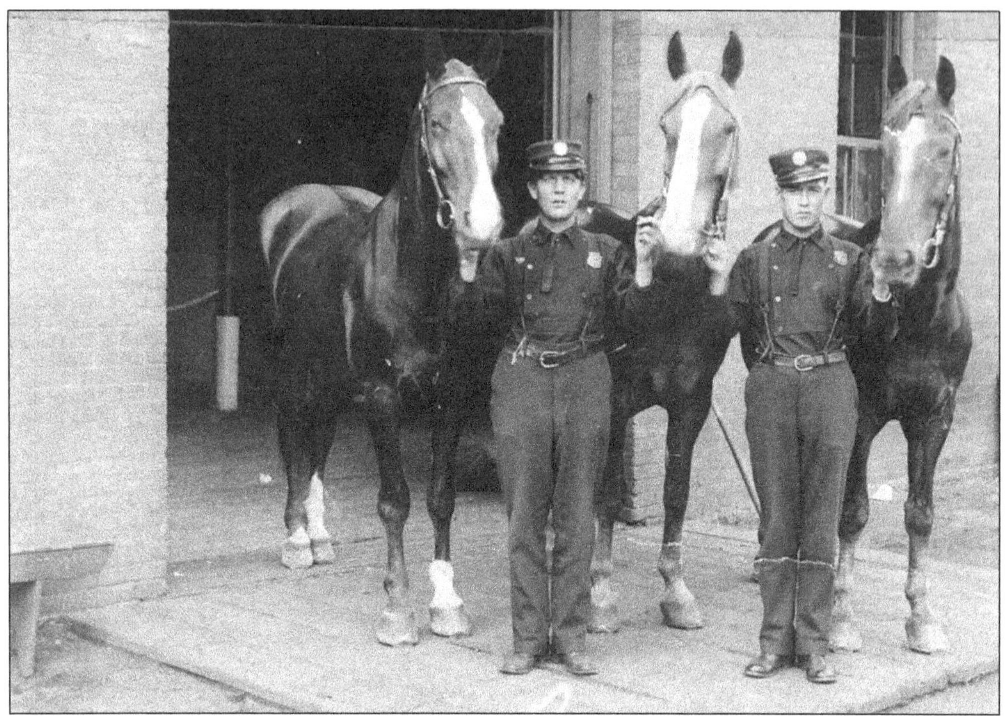

Pictured in the 1890s, two uniformed firemen pose between three horses, which they are holding by the bit rings outside of the doorway of a brick stable. The horses had to be exercised daily in addition to cleaned and checked on. A veterinarian would come by the firehouse weekly or as needed to examine the horses. (Courtesy of the Detroit Historical Society.)

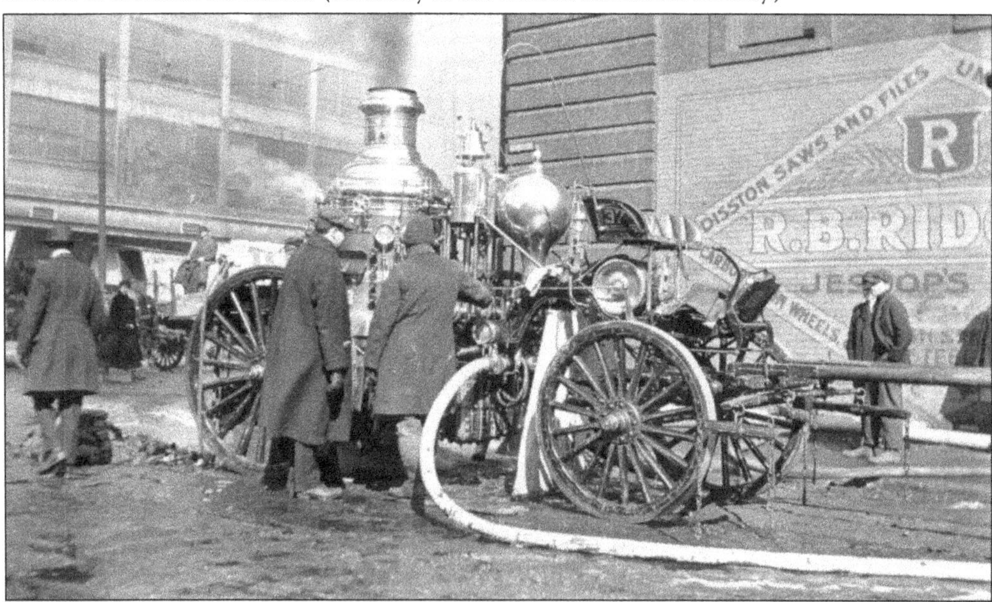

Engine Company No. 37 is shown operating at a fire on First Street, just south of Congress Street. Smoke rises from its stack, and hose lines are charged. Note the horses have been removed and taken to a nearby firehouse or stable after pulling the heavy steamer the required distance. (Courtesy of the Detroit Historical Society.)

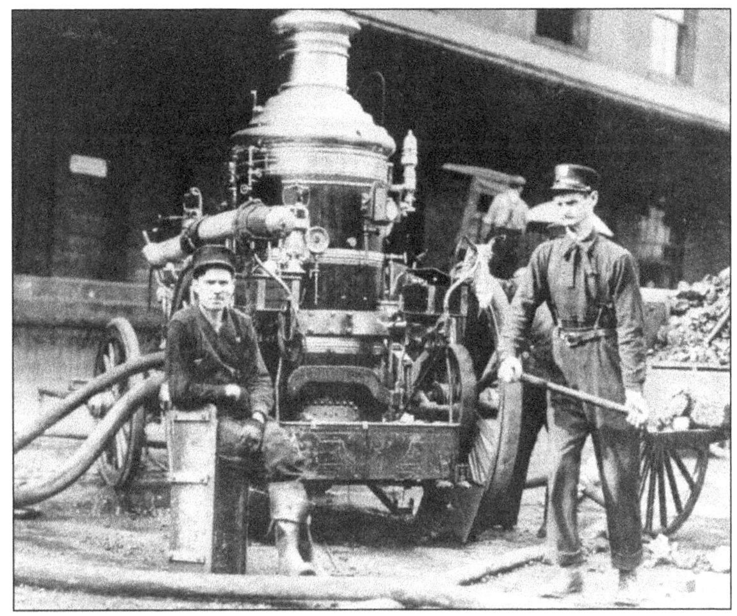

Two firemen working at a fire scene are located at the rear of their operating steam engine. The fireman on the right holds a shovel that he uses to stoke the boiler with coal and or wood for fuel. The other fireman on the left sits atop a box. (Courtesy of the Detroit Historical Society.)

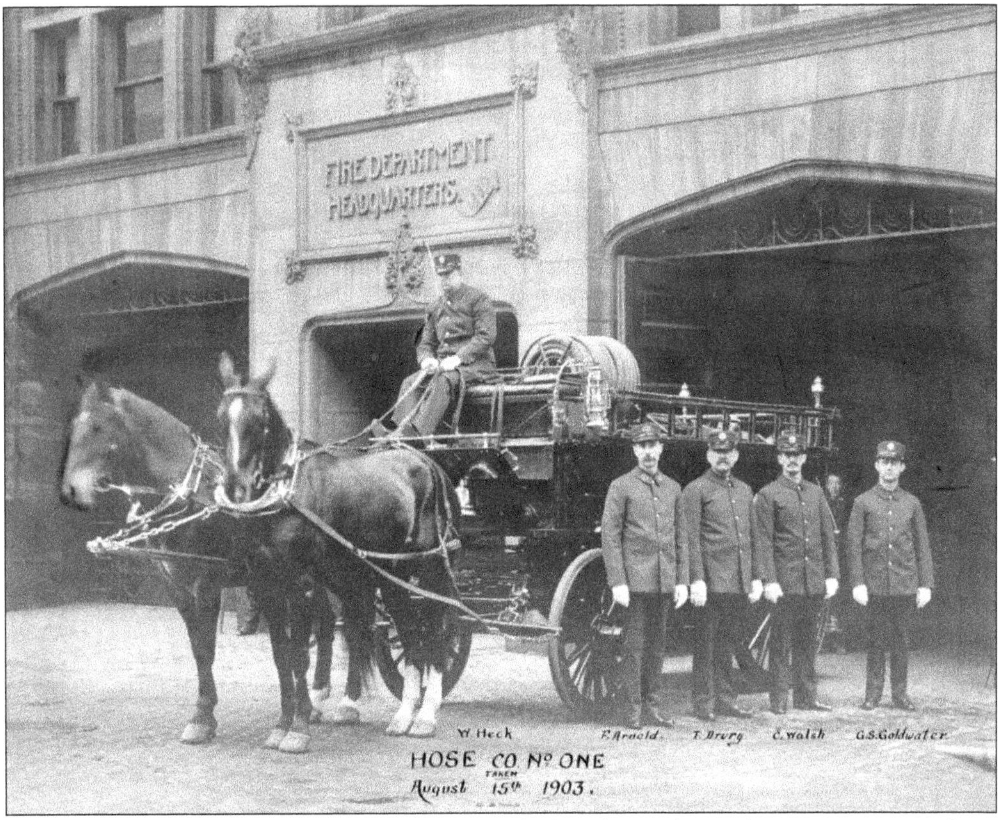

The firemen of Hose Company No. 1 pose with their horse-drawn hose apparatus in front of the entrance of the Detroit Fire Department Headquarters. A stone banner over the door of the headquarters reads, "Fire Department Headquarters 1882." (Courtesy of the Detroit Historical Society.)

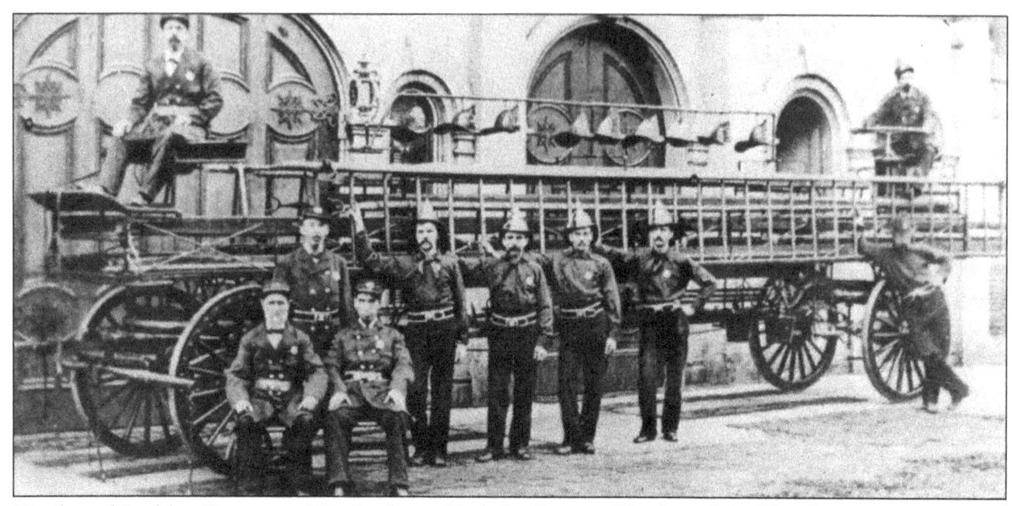

Hook and Ladder Company No. 1, also called the Rescue Hook and Ladder Company, operated out of the headquarters firehouse, located at Larned and Wayne Streets. In this photograph, members show off their horse-drawn apparatus. The men's helmets are hung on a top rail for easy access and placement. At that time, the ground ladders were all a wooden straight-length design; there were no extension ladders. Engine Company No. 1 was also housed there, as well as the offices of the Detroit Board of Fire Commissioners. (Courtesy of Matt Lee.)

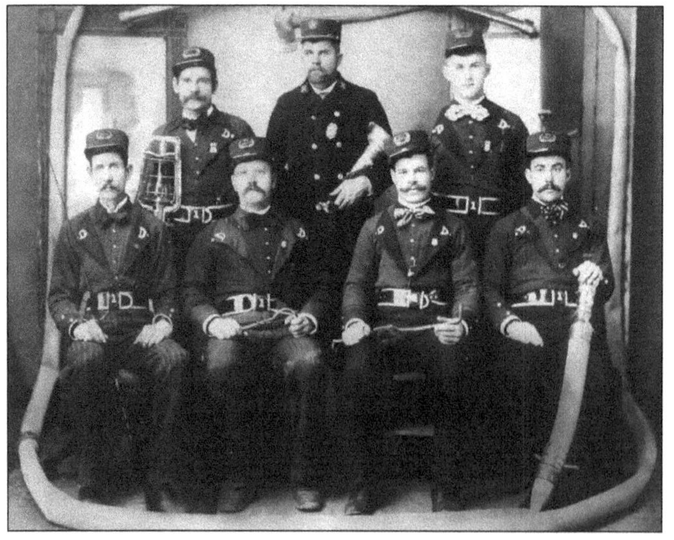

Members of Detroit Fire Department Hose Company No. 1 pose for this photograph. The individuals appear to have various ranks and positions in the company. A battalion chief with a speaking trumpet stands alongside an officer and a fireman, while the driver of the apparatus with the whip in hand and firemen with hand tools are seated. A fire hose outlines the group for a classic touch; a fireman is holding the end of the nozzle. (Courtesy of Matt Lee.)

Members of Engine Company No. 29 pose on a muddy street. A group of young children looks on with interest in this residential neighborhood on Detroit's southwest side. While responding to calls, apparatus like this horse-drawn rig had some problems navigating the wheel tracks made by other wagons until the City of Detroit paved the streets. The horses could have special shoes fitted to their hooves to accommodate the terrain. (Courtesy of Matt Lee.)

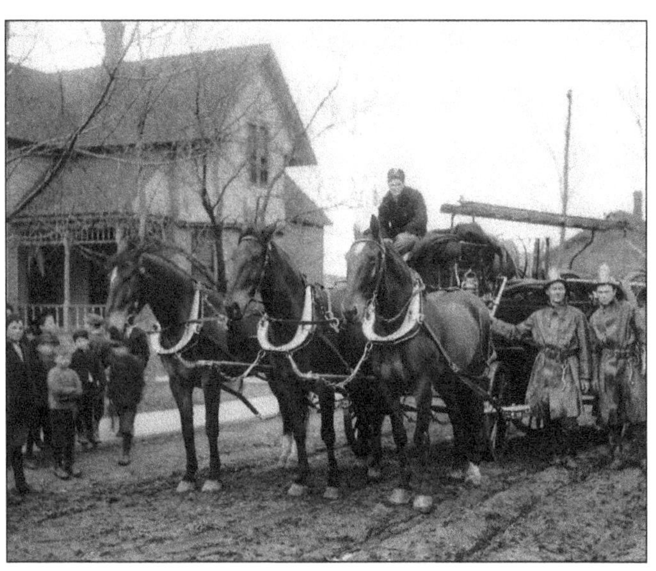

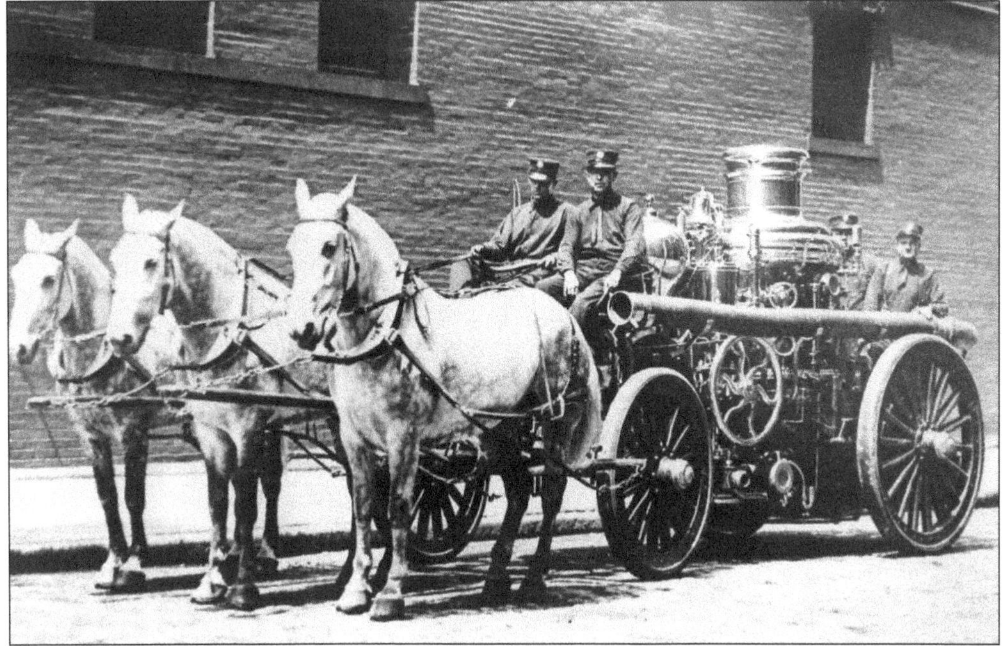

Members of Engine Company No. 1 pose with their steamer. This photograph shows a four-man crew where most photographs showed a three-man crew operating on the apparatus. The breed of horse used in the fire service was called Percheron, a type of draft horse. Horses had to pull the steamer for distances up to 8 to 10 blocks. A short run would take about five to seven minutes. The average horse served anywhere from three to seven years. Engine No. 1 was located at Wayne and West Larned Streets. Wayne Street later was renamed Washington Boulevard. (Courtesy of Matt Lee.)

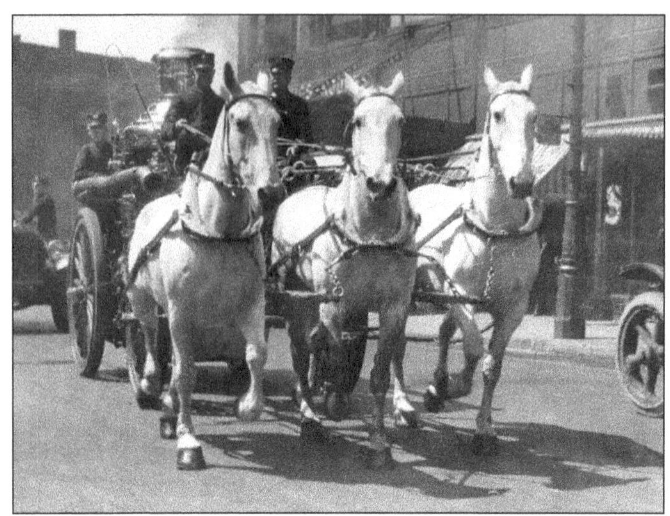

This steam engine is responding to a fire call. The horses would gallop while pulling behind them the steamer that weighed about 7,000 pounds. The fireman riding on the back step would have started a fire in the firebox before leaving the firehouse in order to have the boiler ready once it arrived on the scene to produce pressurized water. The fuel used would consist of coal and excelsior (fine wood shavings). (Courtesy of David Traiforos.)

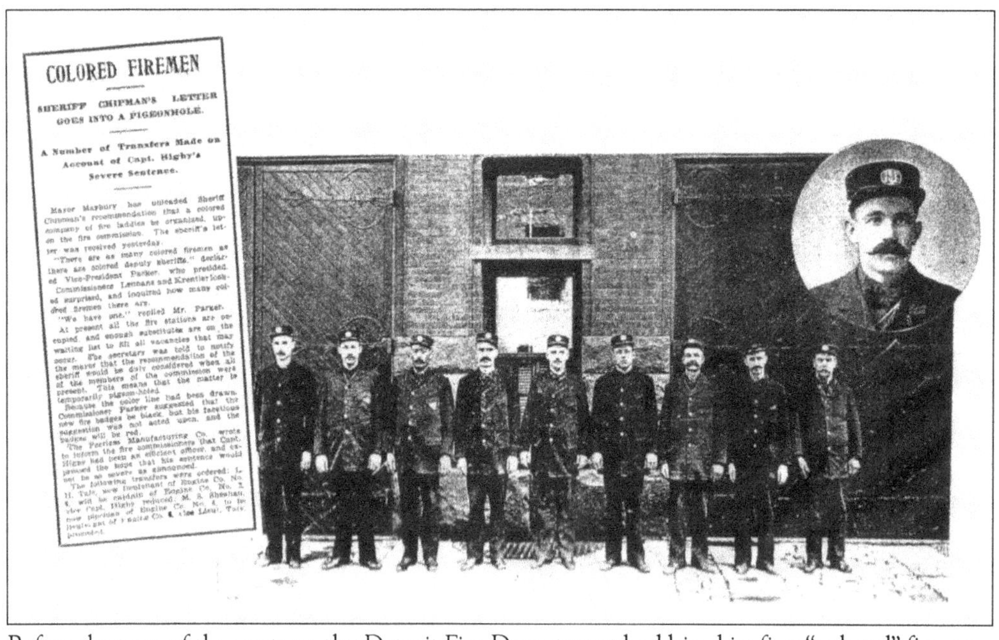

COLORED FIREMEN

SHERIFF CHIPMAN'S LETTER GOES INTO A PIGEONHOLE.

A Number of Transfers Made on Account of Capt. Higby's Severe Sentence.

Before the turn of the century, the Detroit Fire Department had hired its first "colored" fireman. Alfred J. Brown was hired in 1894. He is pictured in this article, third from the left. He left the fire department when offered a job with the US Postal Service. It was later recorded in the history of the Detroit Fire Department that the first African American fireman was hired in the mid-1930s. (Courtesy of the Detroit Firemen's Fund.)

Two

1900s–1920s

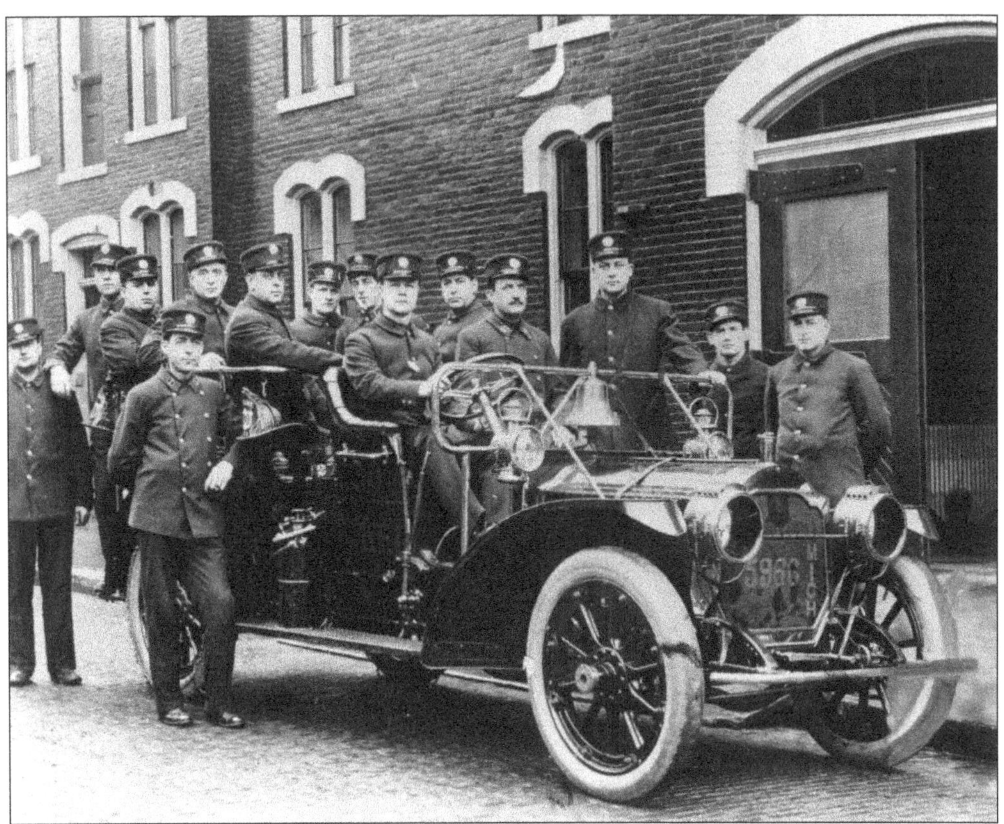

Here, Detroit firemen are seated in or standing beside a rescue car belonging to Engine Company No. 30. The vehicle, made by Packard, is parked on the street in front of a brick building that has a large doorway, probably the fire station. The c. 1910 vehicle is right-hand drive since the steering wheel is on the right side of the car. (Courtesy of the Detroit Historical Society.)

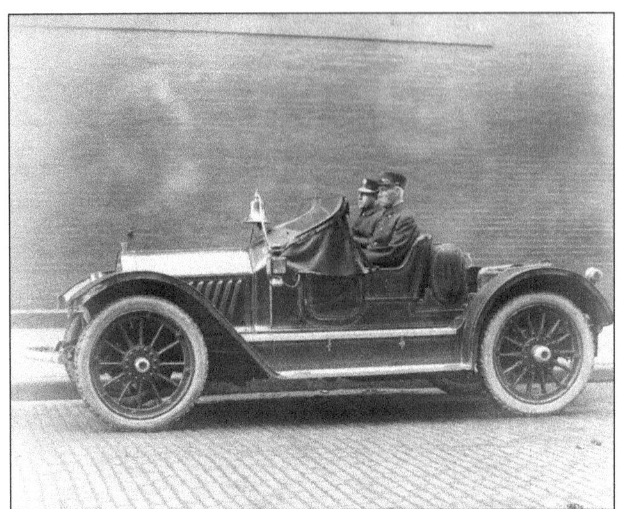

Pictured around 1915, Detroit Fire Department chief James C. Broderick and driver George Appel are seated in an Oldsmobile Autocrat. The symbol of five crossed bugles indicated the fire chief and designated Broderick's rank as chief of department; it is painted on the side of the car's bench seat. (Courtesy of the Detroit Historical Society.)

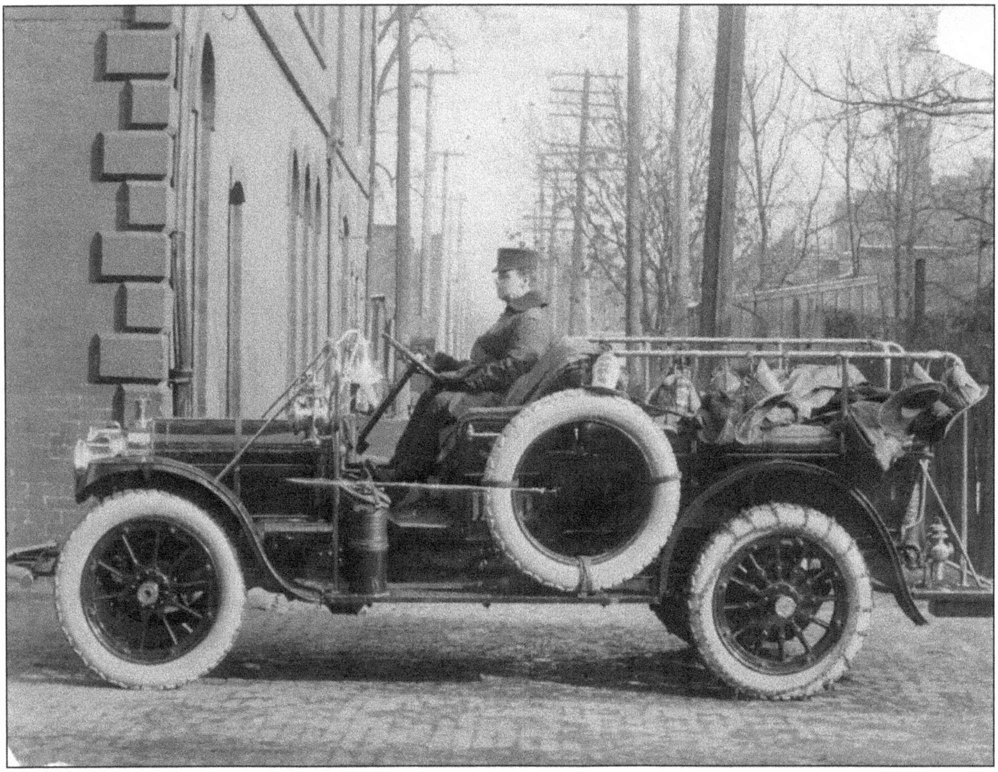

A Detroit Fire Department Packard squad car is parked on a brick driveway near the corner of a building and shows the driver seated on the right of the vehicle. The car is right-hand drive since the steering wheel is on the right side. "Engine No. 30" can be seen on the firemen's helmets hanging from the horizontal bar at the back of the squad car. A second Packard squad car for the department was bought and was in service as old engine No. 30, then located on the east side of Hastings Street between East Larned and Congress Streets, and was quartered with Engine No. 2 in the same station. This unit was delivered on October 18, 1910, to the Detroit Fire Department and was factory serial No. 15957 and department No. 215. The squad car was sent to Engine Company No. 34 (later to become Squad No. 2) in 1912. (Courtesy of the Detroit Historical Society.)

Detroit firemen are responding to a night alarm at the turn of the century. They are getting into bunker pants located next to their bunk bed, thus the name they were given. An officer is shown descending the pole from the third-floor level, which served as his quarters. (Courtesy of Matt Lee.)

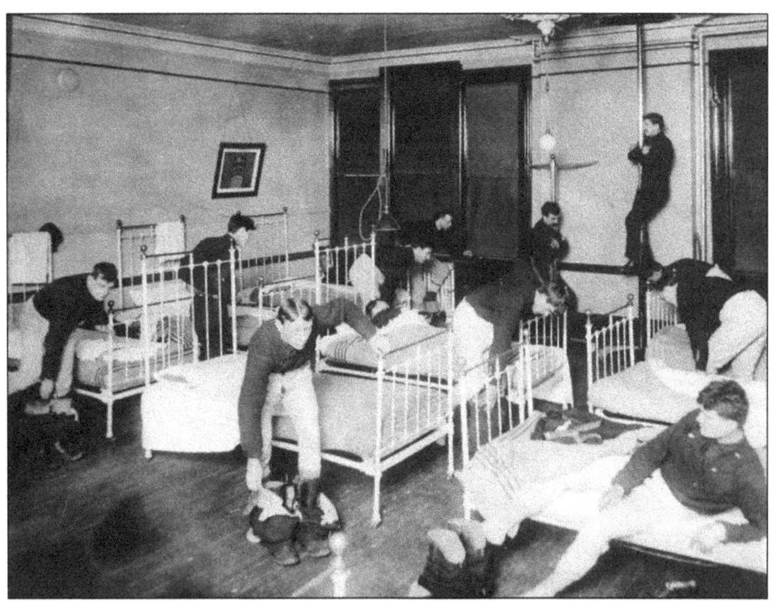

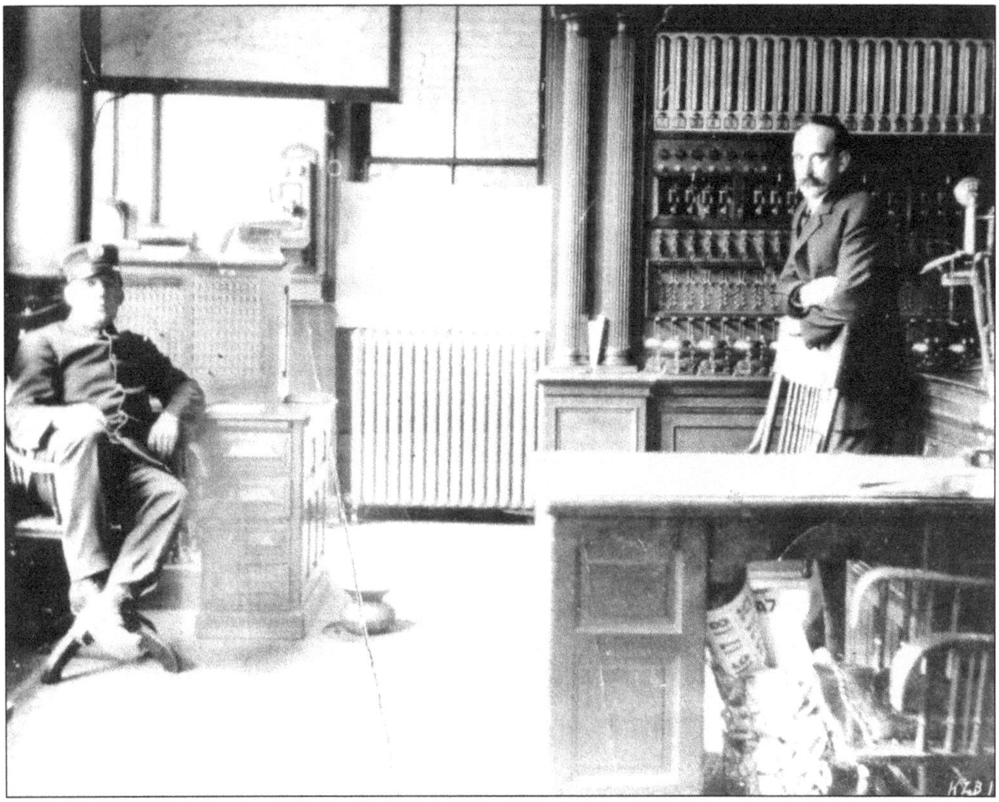

The Central Fire Alarm Office of the Detroit Fire Department was formerly located on Larned Street. This view shows two communications officers posing for the camera. The office was where fire alarm notifications were received with telegraph equipment. Detroit's first workable fire alarm system was installed by the Gamewell Company and was placed in service in November 1870. Note the spittoon and the overflowing garbage can. (Courtesy of Matt Lee.)

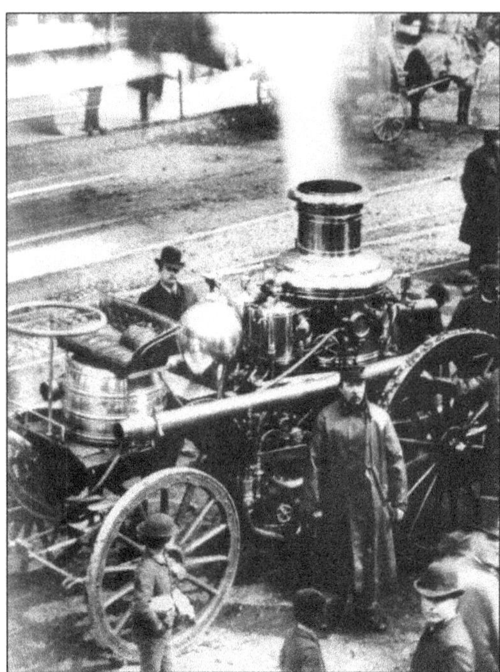

Detroit's first self-propelled steam fire engine was built in Manchester, New Hampshire, by the Amoskeag Manufacturing Company in 1874 as a steam engine and later rebuilt in 1890. This vehicle was now motor driven and did so without horses. Here, it served as Engine Company No. 3. (Courtesy of Matt Lee.)

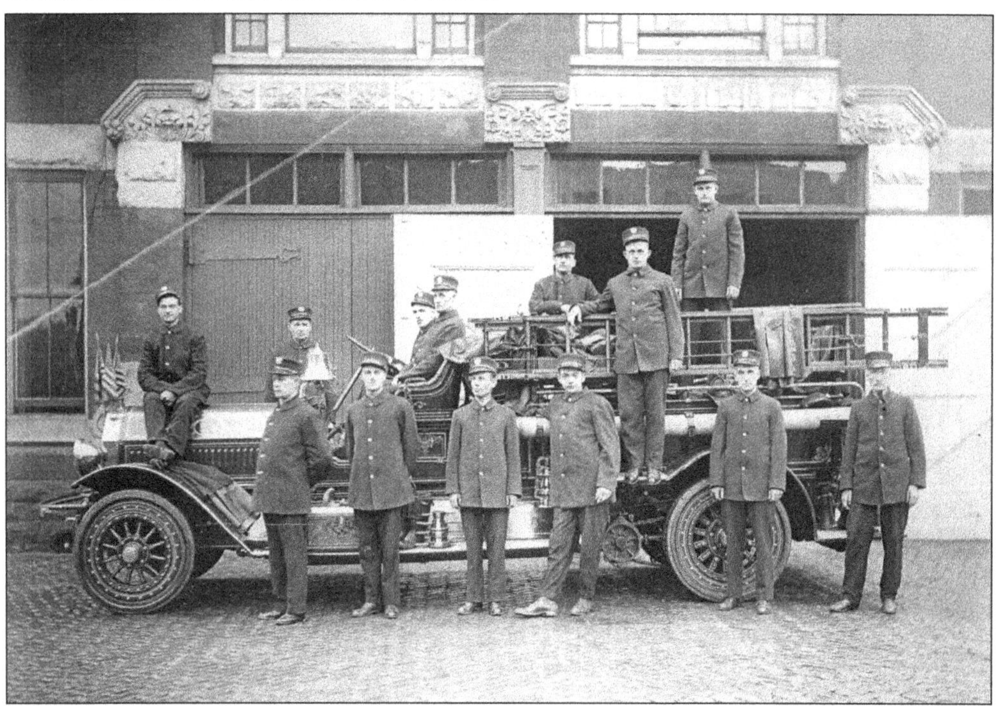

Detroit slowly became known as the Motor City because of the growth in the auto industry, and with its growth came motor-driven fire engines. The new engines started to replace horse-drawn steamers throughout the city of Detroit. Here, firemen pose alongside a new chain-driven vehicle. The new engines had the standard pump and could travel farther distances at faster speeds. They could not only carry hose but also an assorted amount of portable ladders and men. (Courtesy of the Detroit Firemen's Fund.)

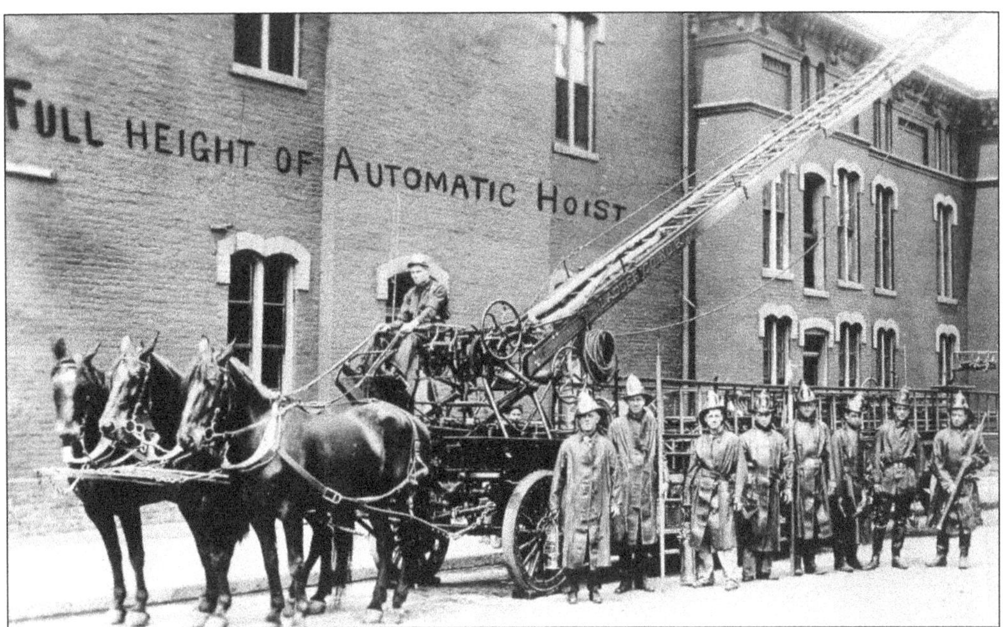

Hook and Ladder Company No. 2 is posing with its horse-drawn automatic hoist ladder. The photograph shows eight members of the company, ranging from captain and lieutenant to firefighters, holding tools wearing different styles of fire coats. The automatic hoist ladder allowed for a ladder to reach people in need of rescue from windows rather than raising ground ladders to reach them. (Courtesy of Matt Lee.)

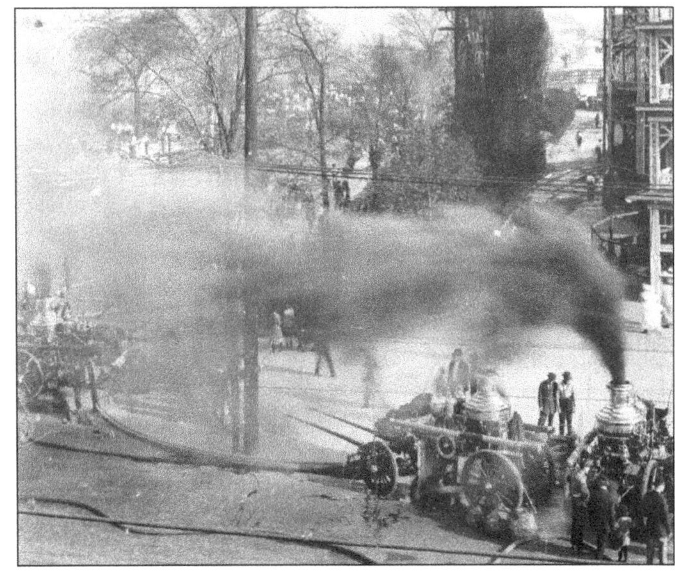
In 1915, a fire broke out on the bridge connecting the city of Detroit to Belle Isle. The wooden bridge sustained serious damage. It spanned a distance of about a half mile. In an effort to fight this fire, many fire companies were needed. Here, several steam engines are working under a full head of steam. The engines were used to pump water long distances to supply water to combat the flames. (Courtesy of the Detroit Firemen's Fund.)

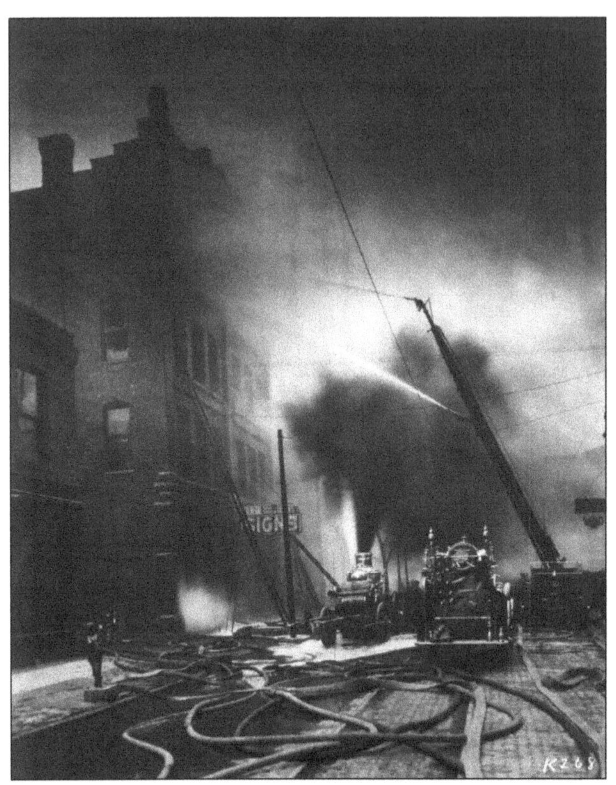

This 1914 blaze at the Federal Electric Sign System Company went to three alarms before being brought under control. The fire occurred at 63 State Street near Washington Boulevard. In front of the building is a steam engine working two hose lines pumping water to the area's first water tower, purchased in 1893. Also in the foreground is Hose Company No.1. (Courtesy of Matt Lee.)

A Detroit Hose Wagon returns to the firehouse after having responded to a fire call. The driver of the hose wagon has directed the horses to drive up the apron into the firehouse. A fireman with a hose washes down the wagon and wheels before going into the firehouse to turn the wagon around on the apparatus floor facing out the door. (Courtesy of the Detroit Firemen's Fund.)

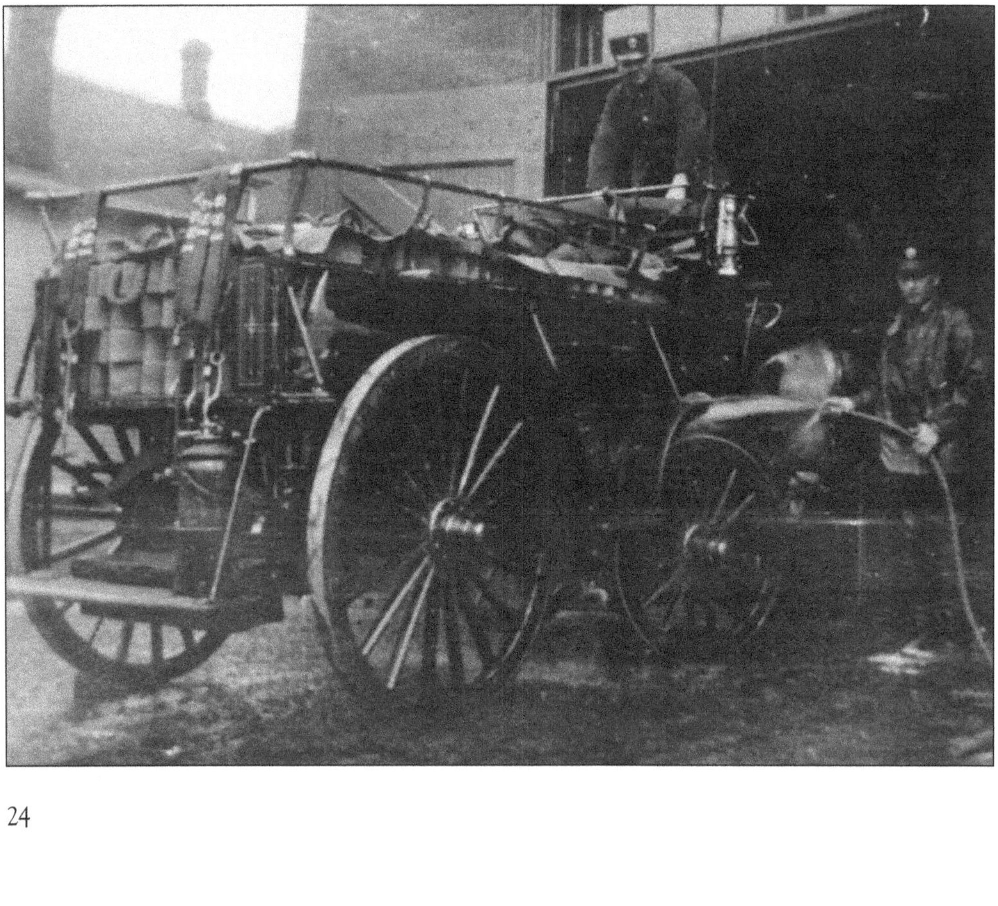

Captain David Baxter of Engine Company No. 17 poses in August 1911 near the watch desk of the firehouse located at Second Avenue and Burroughs Street, which is still open and in service today. Engine No. 17 was and is still in the same firehouse with Ladder Company No. 7, which is also still in service today serving the New Center area of Detroit by the Fisher Building. (Courtesy of the Detroit Firemen's Fund.)

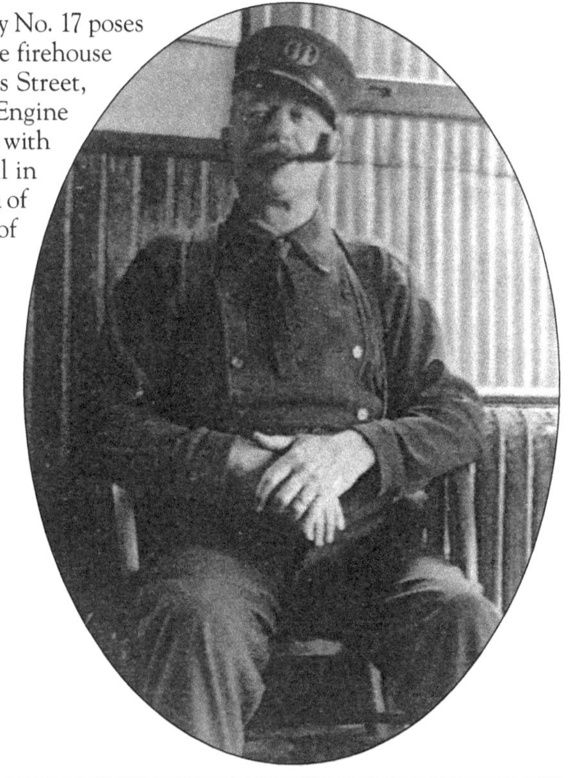

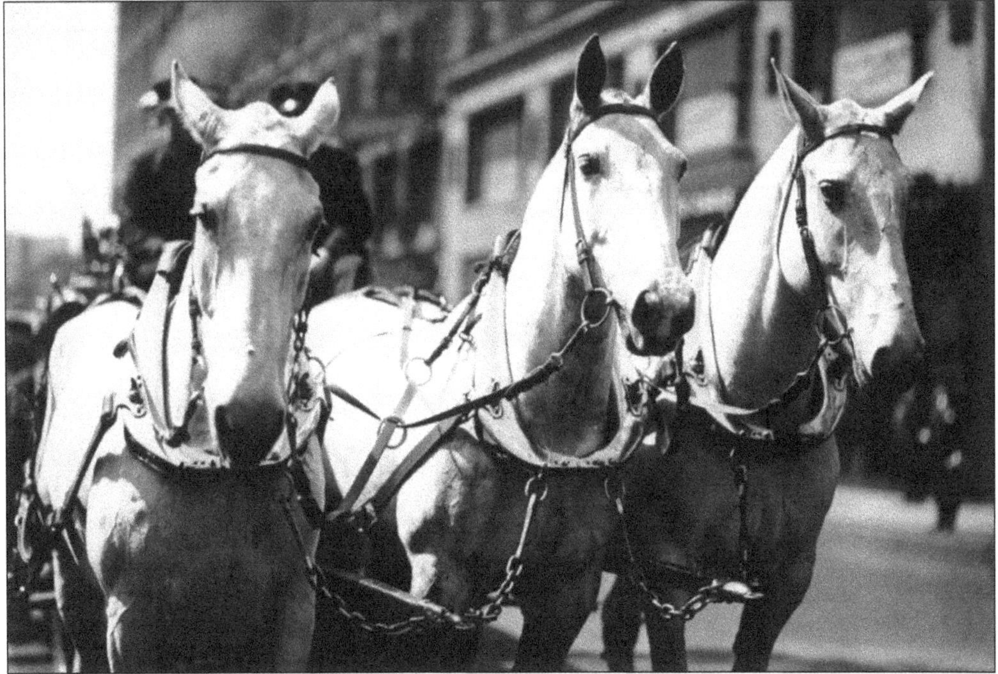

A team of three Detroit Fire Department horses fully bridled poses for before their last engine run through downtown Detroit. This was part of the ceremonial final run of the horse-drawn Detroit Fire Department apparatus on April 10, 1922. (Courtesy of the Detroit Historical Society.)

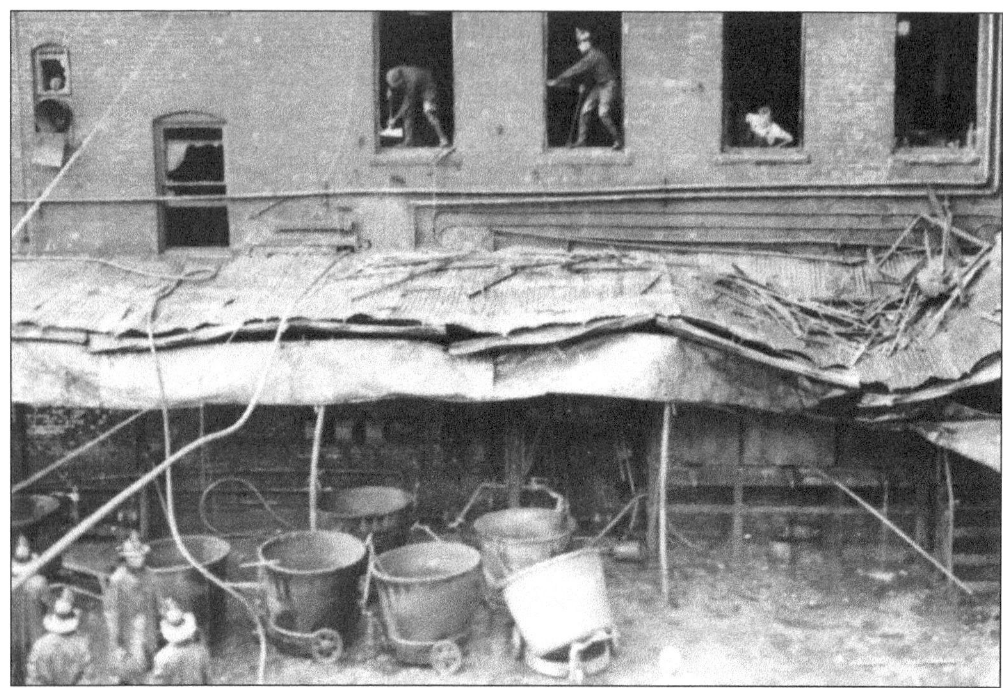

Around 1929, Detroit firemen work on performing overhaul after a fire in this commercial building at Twelfth Street and West Jefferson Avenue. They are also seen working in windows to pull apart the frame work and check for any hot spots left over to avoid a rekindle. (Courtesy of the Detroit Firemen's Fund.)

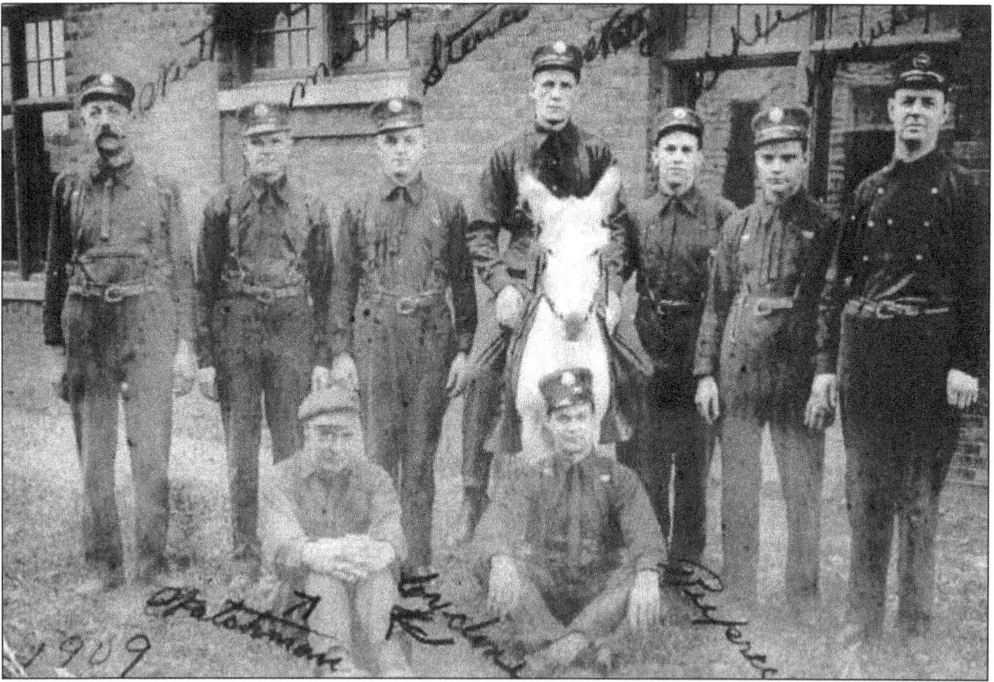

Back in 1909, members of a Detroit fire company pose for a photograph. Note that one fireman is sitting on a pony. (Courtesy of the Detroit Firemen's Fund.)

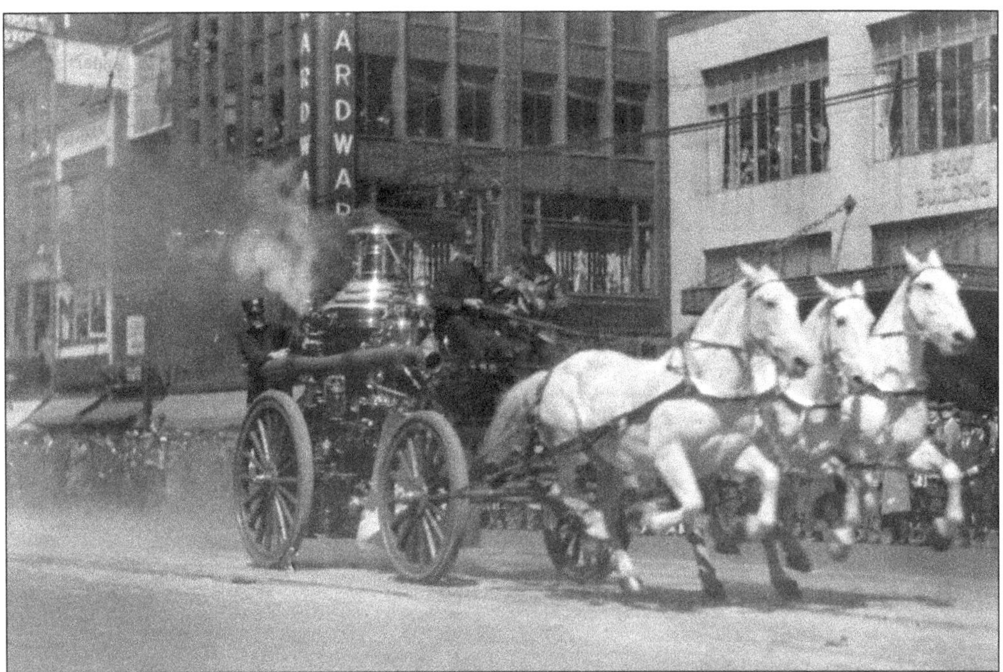

The ceremonial final run of a horse-drawn Detroit Fire Department apparatus was completed by Engine No. 17 on April 10, 1922. Taken from the west side of Woodward Avenue south of Grand River Avenue, this photograph shows three firemen riding on a fire engine being drawn by a team of three horses. The Shaw Building and Rayl's Hardware stand in the background. Spectators line the sidewalk and peer through the windows overlooking the street. (Courtesy of the Detroit Historical Society.)

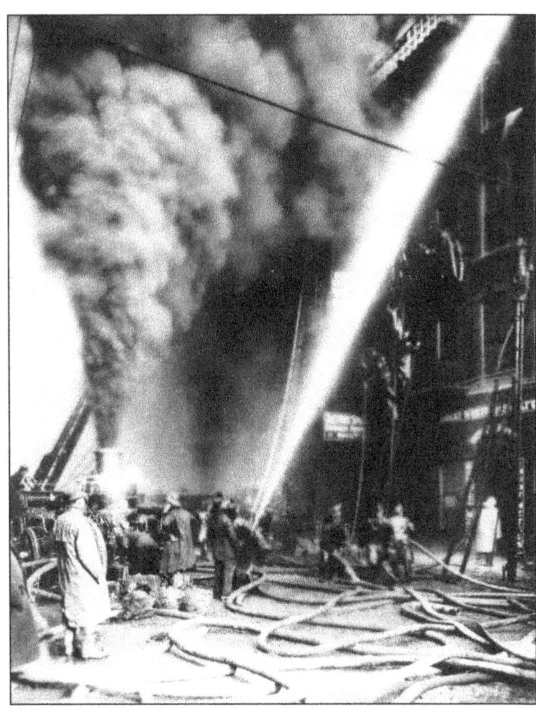

On February 5, 1911, a fire broke out at Bates & Woodbridge. The fire in a commercial building spread to other buildings. The blaze escalated to a multiple-alarm fire. Many steamers had to be used to supply hose lines for firemen to fight this blaze and bring it under control. Here is a classic look at a steamer working. Many times at the scenes of fires, the steamer's smoke was worse than the building's smoke since it stayed low to the ground and obscured visibility. (Courtesy of the Detroit Firemen's Fund.)

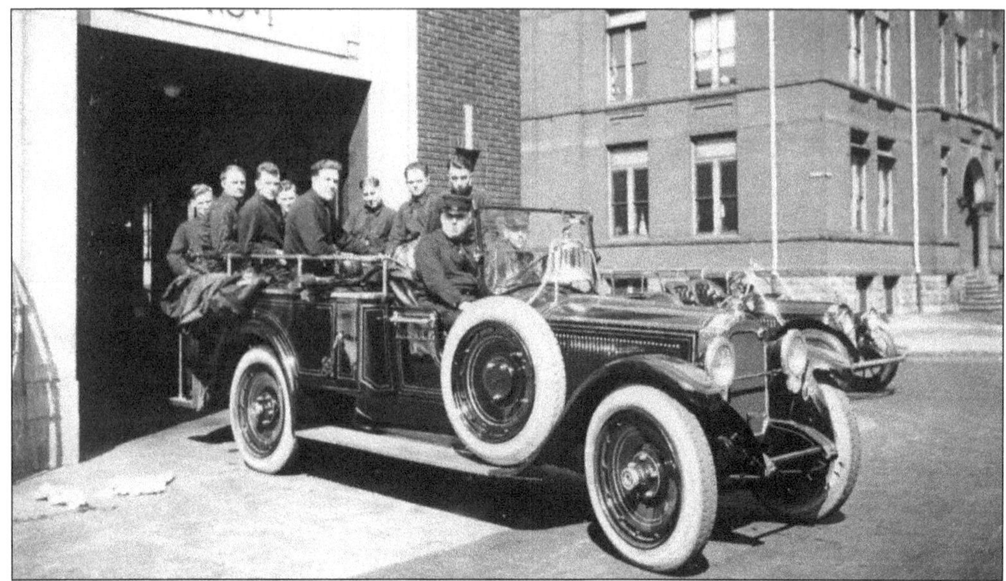

Rescue Company No. 1 is shown with its 1922 Packard squad car. This vehicle had solid-white rubber tires along with a spare tire. Ten members are shown riding on this rig. Responding to calls back then exposed the members to all the elements. This firehouse was erected in 1918 and joined the quarters of Engine Company No. 2 located at East Larned and St. Antoine Streets. Today, the Packard plant, what was once one of the busiest car and truck manufacturers in Detroit, sits vacant. (Courtesy of the Detroit Firemen's Fund.)

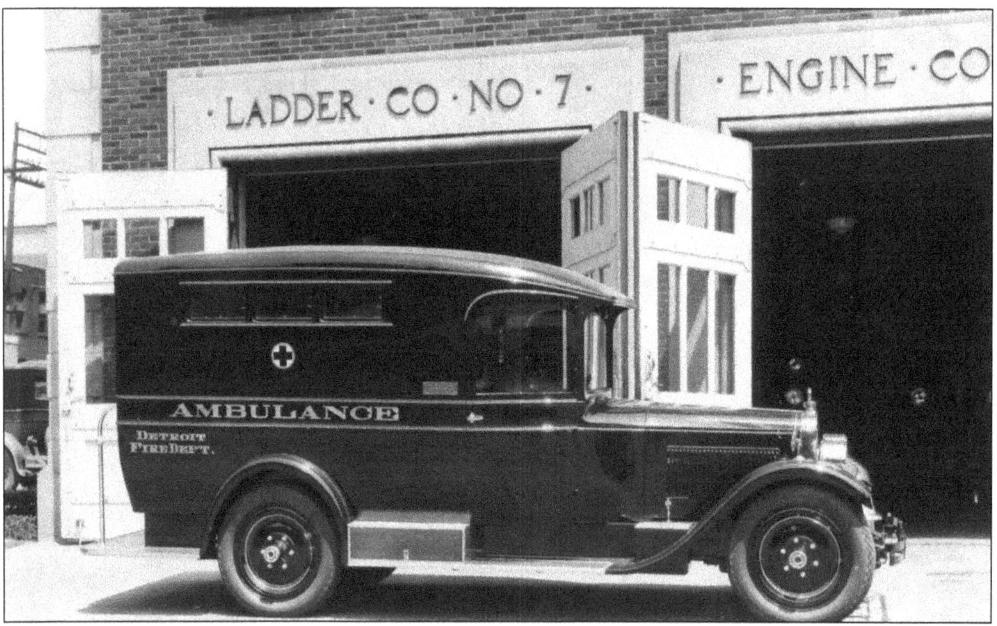

Detroit's first ambulance, a 1927 Packard, was presented to the fire department by Mrs. Lydia Mendelssohn, whose son was a member of the board of commissioners. This was the first introduction to what today is called EMS. Over the years, the DFD has had many different ambulances staffed by firemen. Up until the 1970s, when there was a dramatic increase for EMS service, the DFD provided the separate medical service. (Courtesy of the Detroit Firemen's Fund.)

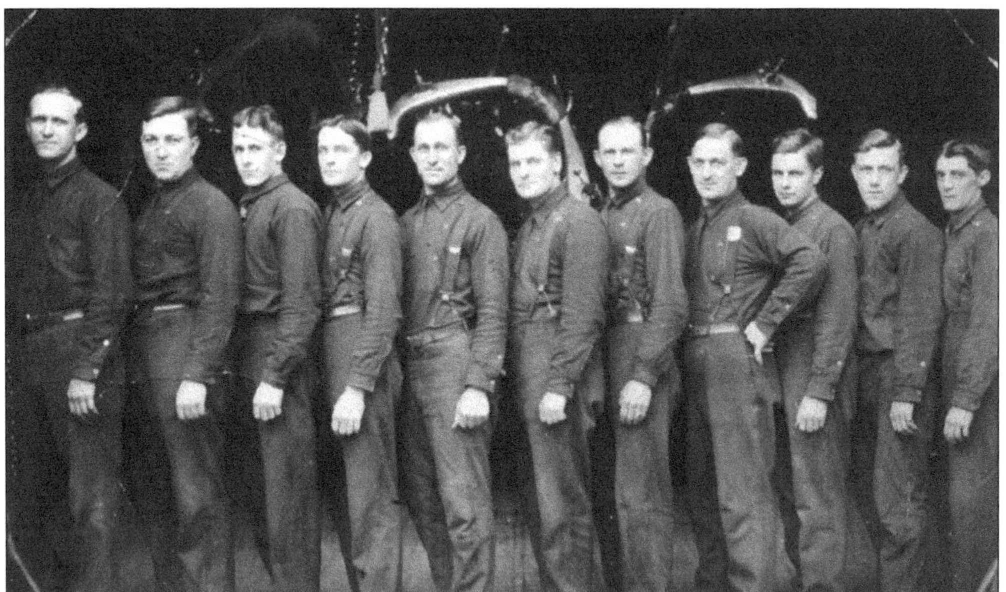

Members of Ladder Company No. 6 are seen standing in front of their firehouse located at East Congress Street and Jos Campau Avenue. Note the horse harnesses hanging where the horses would position themselves, aided by a fireman, to be hitched to the apparatus. The firehouse was opened in 1894, and No. 6 was housed with Engine Company No. 19. Ladder Company No. 6 is still in service today, serving just east of the downtown area. (Courtesy of the Detroit Firemen's Fund.)

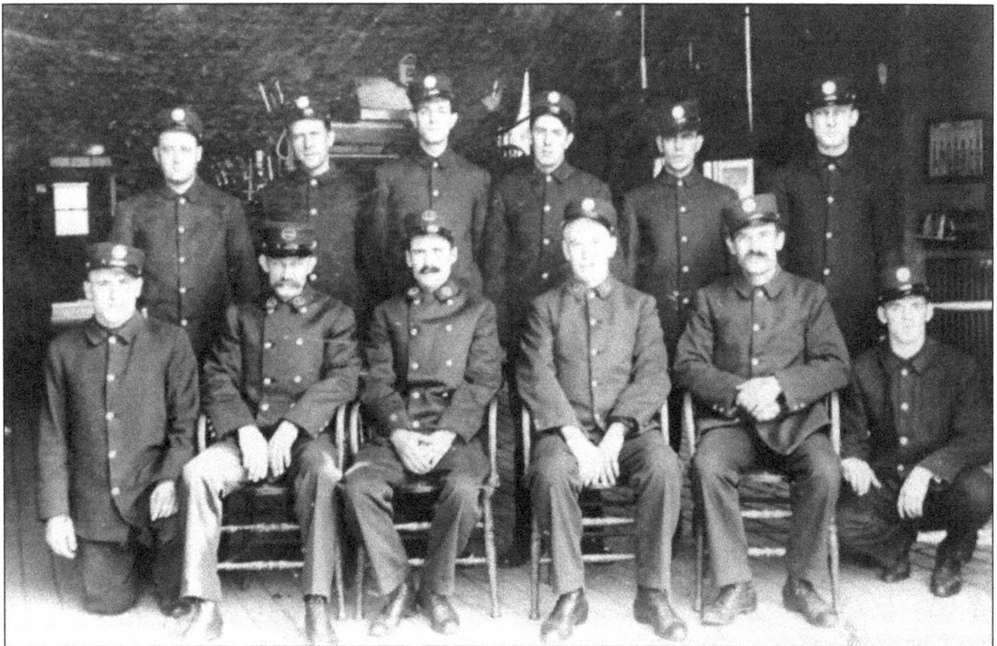

Members of Ladder Company No. 22, including the captain and lieutenant, pose with their apparatus in their firehouse located at McGraw Avenue and Martin Road. The old wooden floor has been replaced with a cement floor; the old radiators in the background are still present and provide heat to that portion of the firehouse. (Courtesy of the Detroit Firemen's Fund.)

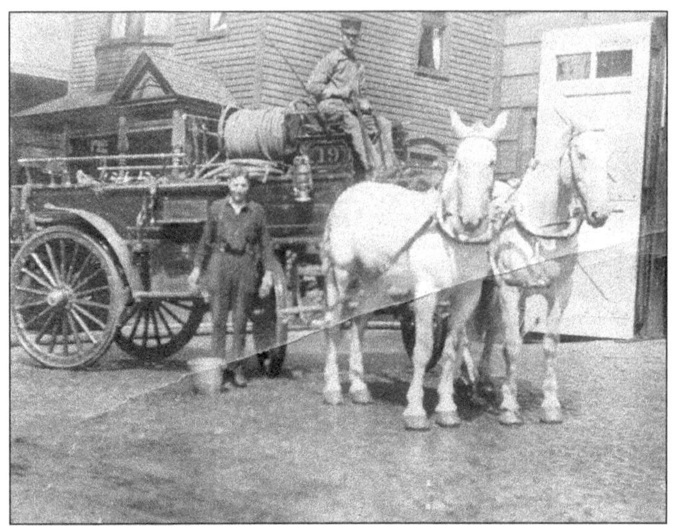

A rare c. 1910 photograph shows Engine Company No. 19 in front of its firehouse. Here, out on the apron, a fireman has just finished wiping down the horses. He has a bucket and towel in hand, and the rig is getting ready to reposition back into quarters. Engine No. 19 was located at East Congress Street and Jos Campau Avenue and shared the firehouse with Ladder Company No. 6. (Courtesy of the Detroit Firemen's Fund.)

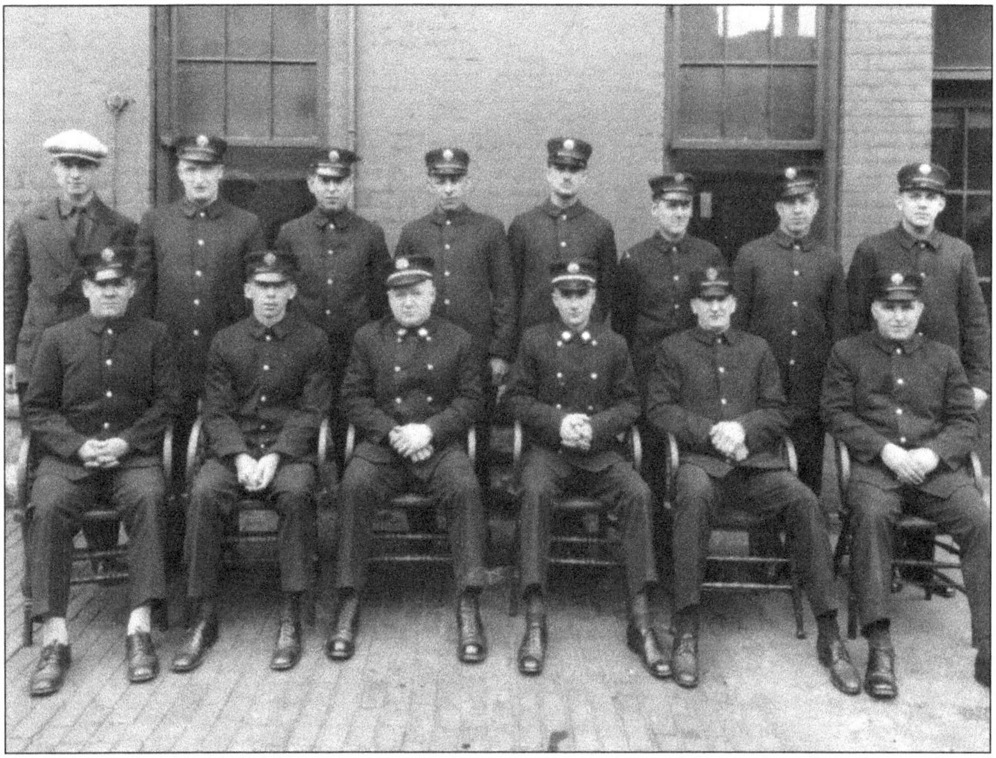

Here, members of Engine Company No. 12 and Ladder Company No. 9 pose out in front of their firehouse around 1928, located at Twelfth and Merrick Streets. The firehouse was built back in 1886 when Engine Company No. 12 was established. In 1894, Ladder Company No. 9 was established and moved in. During lunch on January 16, 1947, instead of delivering diesel fuel, a gasoline fuel truck made a delivery to Engine Company No. 12's firehouse and caused an explosion resulting from the gasoline vapors. The explosion destroyed the firehouse and killed two Detroit firemen and injured 13 others. Several of the men were blown across the street. Firemen Charles Perish and Paul Reiner, both assigned to Ladder Company No. 9, died as a result of injuries. Today, both companies are closed and out of service. (Courtesy of the Detroit Firemen's Fund.)

Three

1930s–1940s

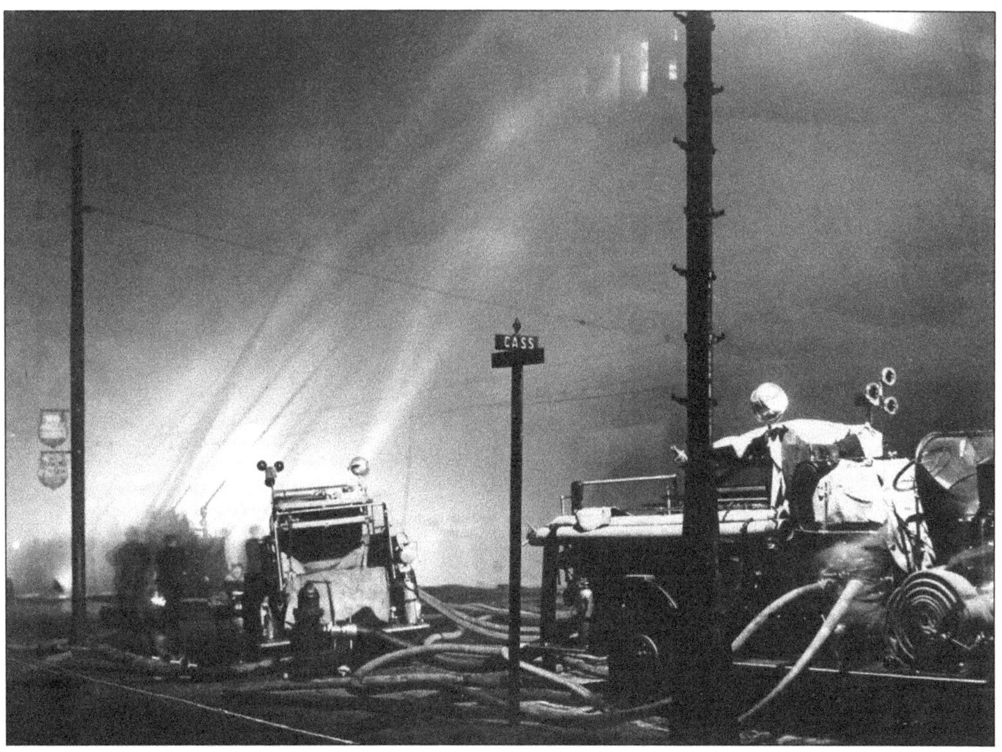

A five-alarm fire took place at the Advanced Glove Company, located on West Jefferson Avenue between Griswold and Shelby Streets on May 10, 1937. A great deal of hose was used to supply many master stream devices that were used. (Courtesy of Matt Lee.)

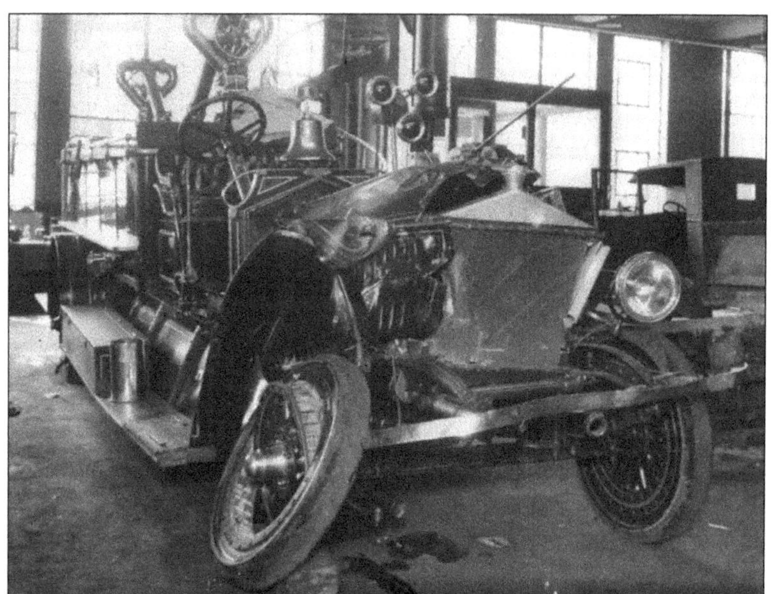

In August 1936, it was reported that Fireman Joseph Hallman was killed when two fire trucks collided on the way to a fire. Company No. 1's Ahrens-Fox high-pressure, open-cab vehicle sustained significant damage. (Courtesy of Matt Lee.)

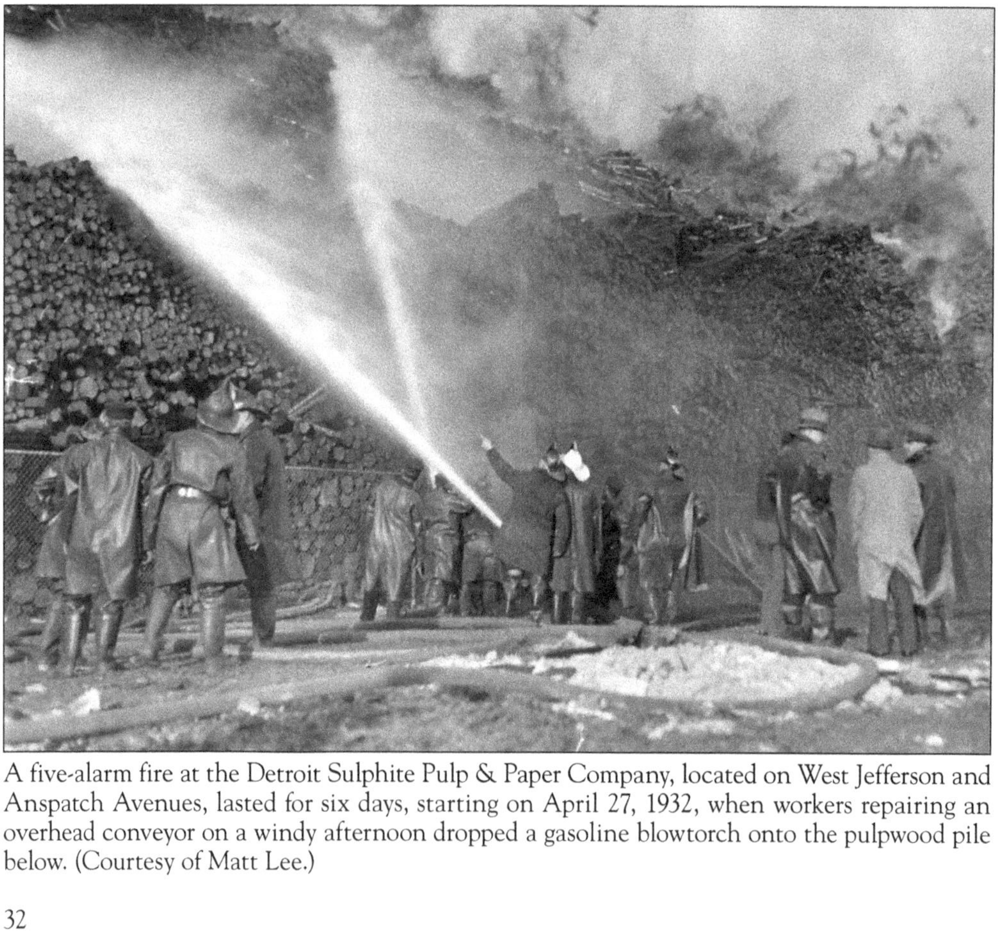

A five-alarm fire at the Detroit Sulphite Pulp & Paper Company, located on West Jefferson and Anspatch Avenues, lasted for six days, starting on April 27, 1932, when workers repairing an overhead conveyor on a windy afternoon dropped a gasoline blowtorch onto the pulpwood pile below. (Courtesy of Matt Lee.)

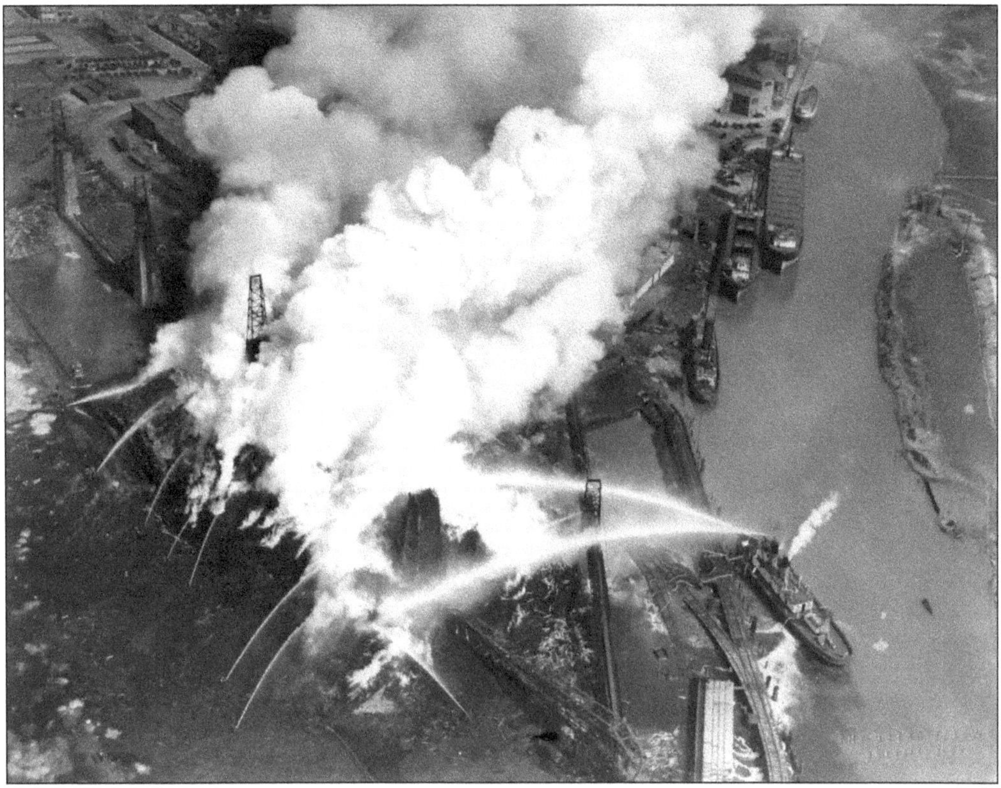

This aerial photograph, taken from the *Detroit News* autogiro, shows Engine Company No. 16's fireboat *John Kendall* working turret guns and hand lines from the Zug Island channel at the Detroit Sulphite Pulp & Paper fire. Other master stream devices are shown being used. Heavy rainfall over several days did help firemen bring this fire under control. (Courtesy of Matt Lee.)

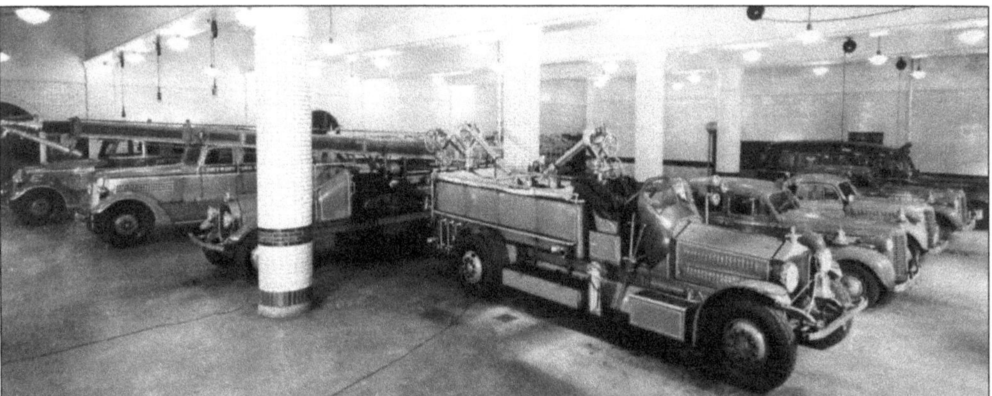

Here is an interior view of Detroit Fire Department Headquarters, located at Larned Street and Washington Boulevard. This firehouse had two sides to the building. One side faced out onto Larned Street with five overhead doors for five vehicles. Three overhead doors faced out onto Washington Boulevard for three vehicles. Headquarters housed Engine Company No. 1's 1936 Seagrave; Ladder Company No. 1's 1937 Seagrave; Tower Ladder Company No. 1's 1924 Seagrave; a 1922 Ahrens-Fox high-pressure wagon; several cars for officers, like two 1938 Buicks, a 1937 Chevrolet, and a 1937 Cadillac; and an ambulance. (Courtesy of Matt Lee.)

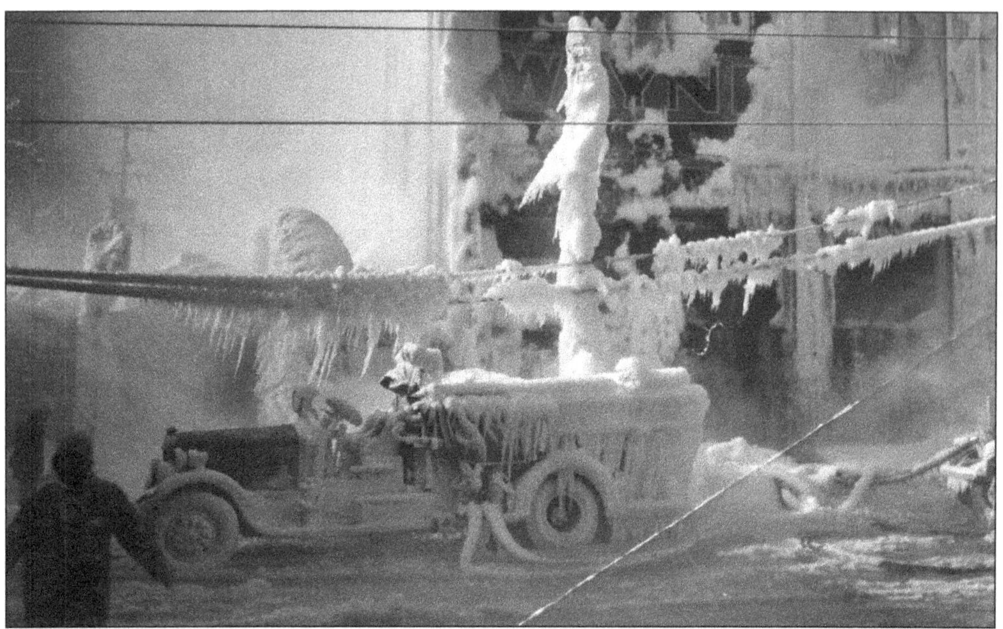

This February 1936 close-up is of an ice-covered Engine Company No. 10 pump in temperatures recorded at 14 degrees below zero at an extra-alarm fire at the Michigan Feed and Grain Company at Sixteenth Street and Grand River Avenue. The vehicle, a 1929 open-cab Mack, had to be towed back to the repair shop by a fire department wrecker, thawed out, and returned to service the next day. (Courtesy of Matt Lee.)

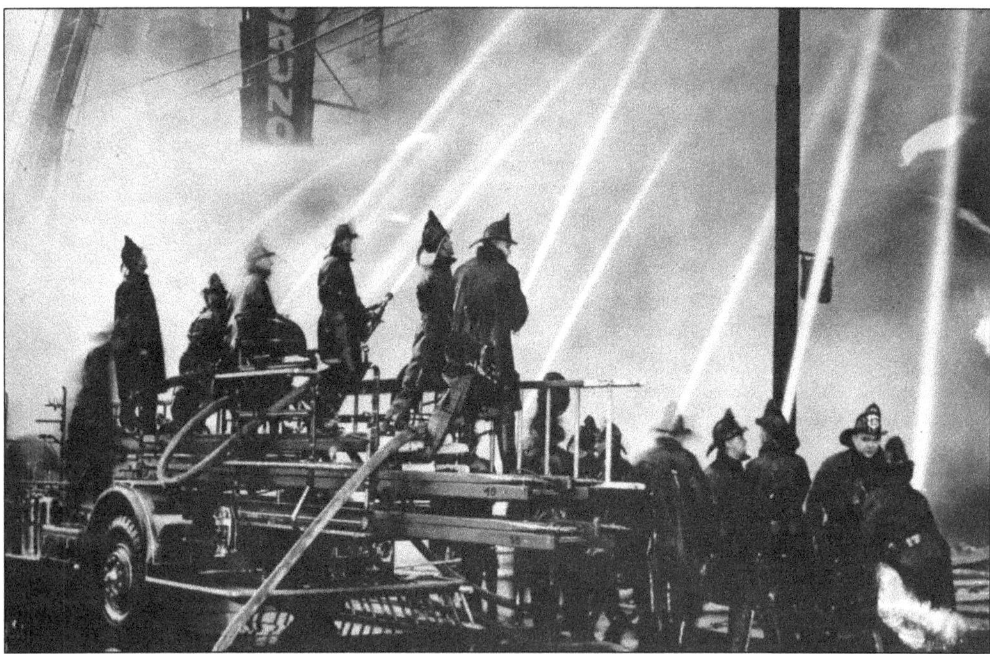

This vantage point of fire attack was captured at a multiple-alarm fire when an aerial device was parked in front of the building on fire. Engine companies stretched their hose lines to this gathering point and directed eight streams of water into the building. Here, the hose lines, pointed into the fire, create a massive attack to extinguish the flames. (Courtesy of the Detroit Firemen's Fund.)

On February 25, 1940, a Detroit fireman, Joseph R. Schneider of Ladder Co. No. 22, died at this four-alarm fire at the Ray Furniture Store, located at Michigan and Livernois Avenues, when he fell through the floor. Master stream devices were used at ground level, and a wooden ladder had flowing water coming from its aerial pipe to extinguish the blaze. (Courtesy of Matt Lee.)

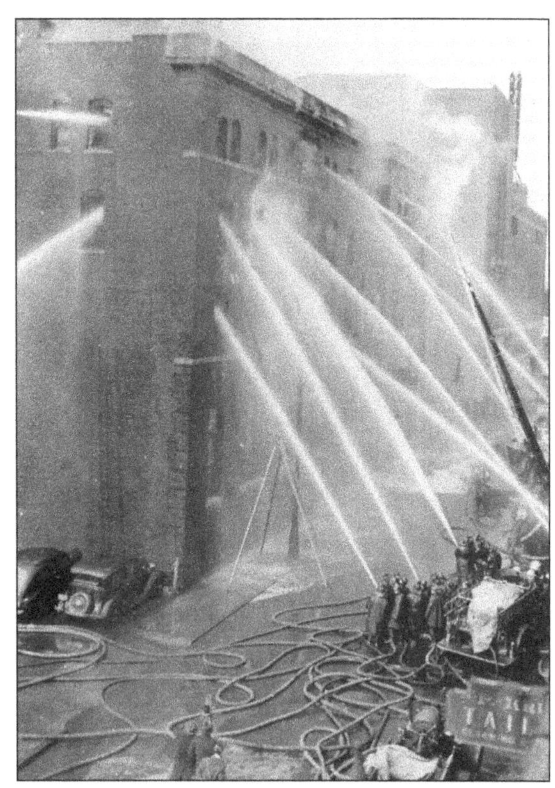

In 1927, Engine Company No. 56, located at Ryan Road and Hildale Avenue, was established. Members pose in front of their Ahrens-Fox along with their firehouse mascot. Engines, ladders, squads, and chiefs were numbered in sequential order. Engines were Nos. 1–60, ladders were Nos. 1–31, squads were Nos. 1–9, and chiefs were Nos. 1–12. (Courtesy of the Detroit Firemen's Fund.)

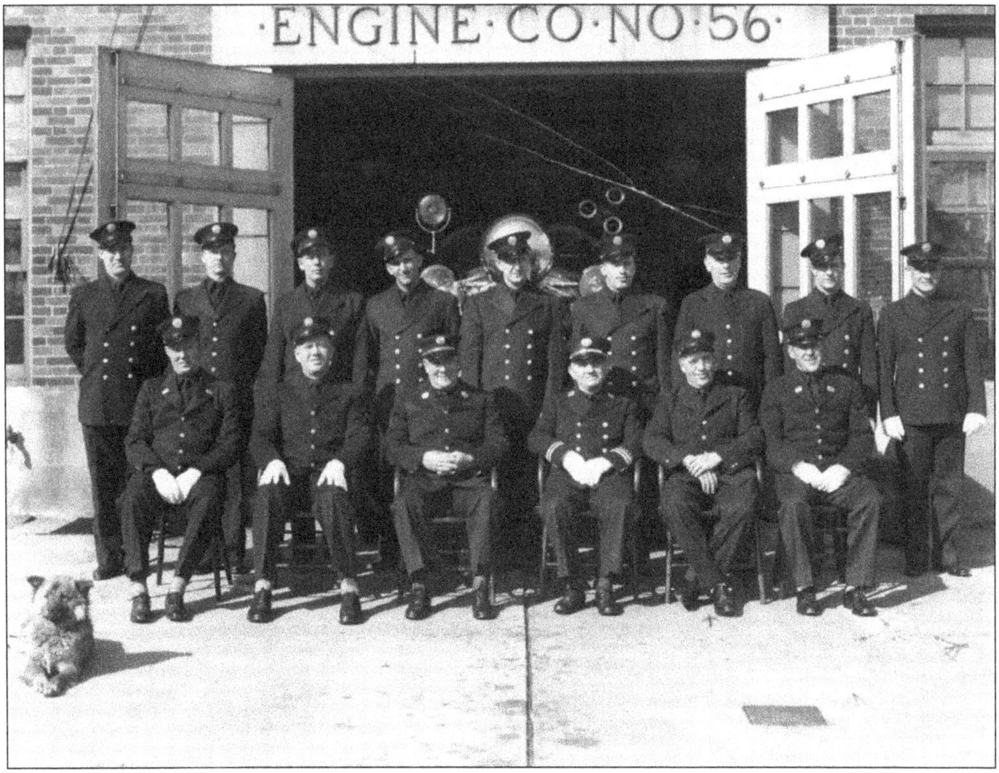

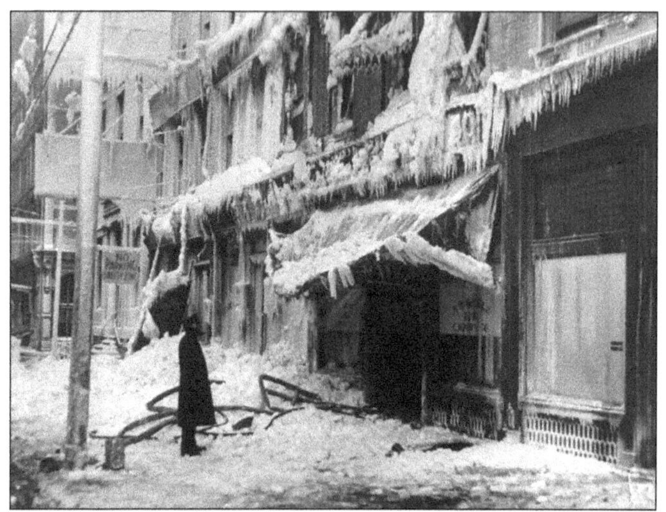

Cold weather fires took their toll on firemen and the buildings that burned down. Here, a fire marshal inspects the ruins of the P.W. Beals Candy Company on West Jefferson Avenue. When the temperature drops below zero, the ice that forms on buildings create spectacular ice castles the day after a fire. But, the ice also creates hazardous conditions for firemen working around these weakened structures. (Courtesy of the Detroit Firemen's Fund.)

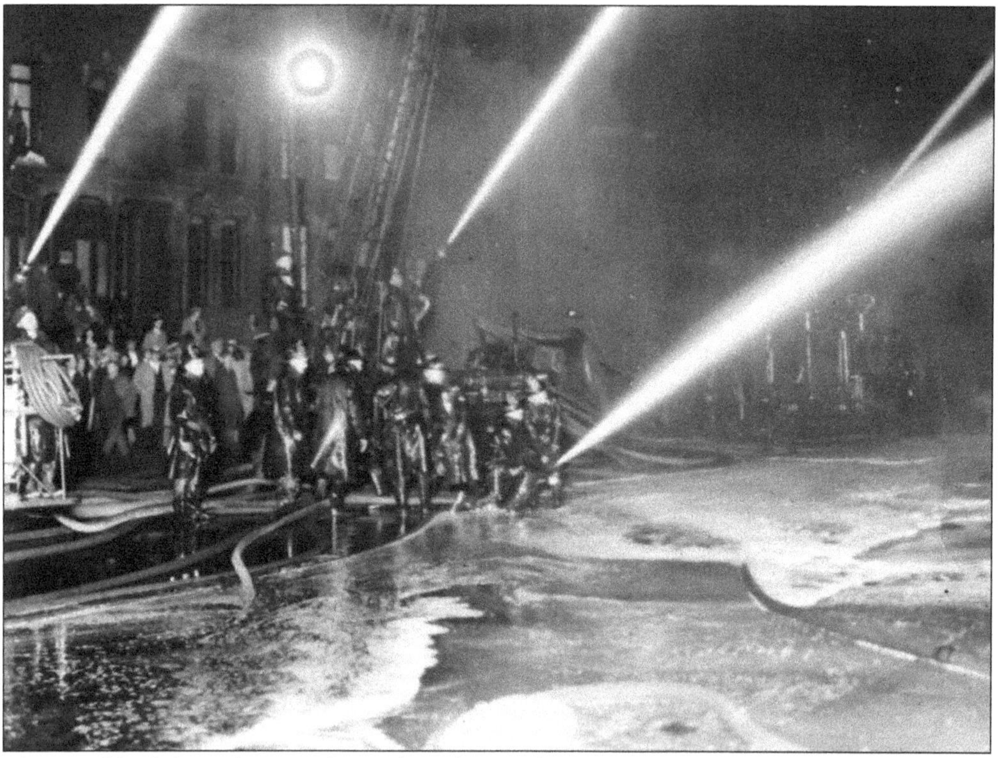

The art of firefighting has not changed much over the years. This fiery scene demonstrates how today's term "defensive operation" was conducted. Master streams, both hand-held and from elevated aerial devices, are directing water into the Ivory Storage Company building around 1930. (Courtesy of the Detroit Firemen's Fund.)

Chief of department Walter Israel observes operations at a five-alarm fire at the Gemmer Steering Wheel Manufacturing Company, located at Sixteenth Street and Merrick Avenue. (Courtesy of Matt Lee.)

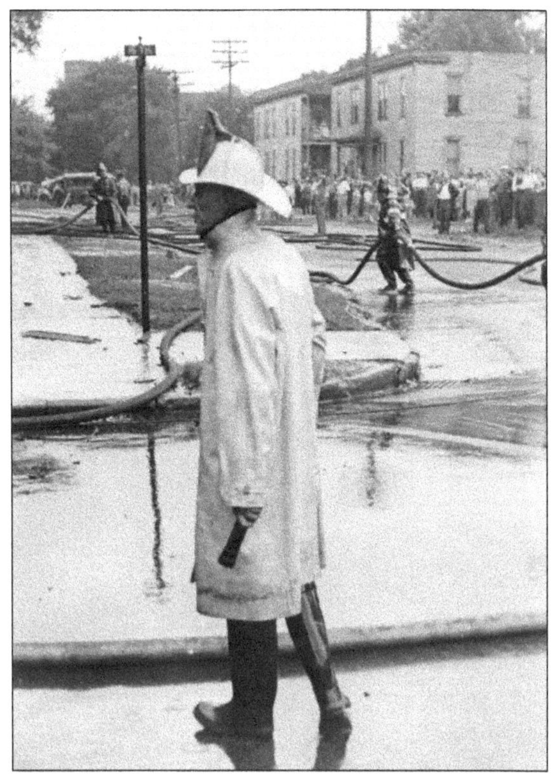

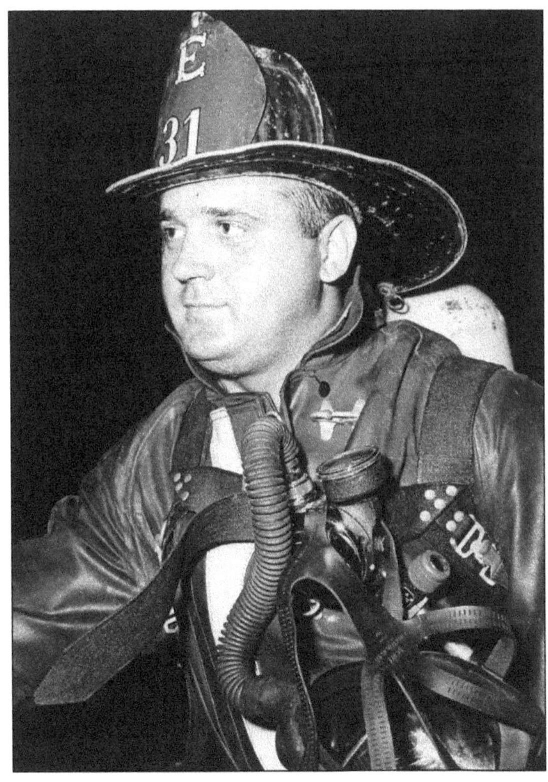

Fireman George Delich of Engine Company No. 31 poses with, at the time, state-of-the-art safety equipment. He has an aluminum helmet, a rubber coat, and a 45-pound, self-contained breathing apparatus. (Courtesy of Matt Lee.)

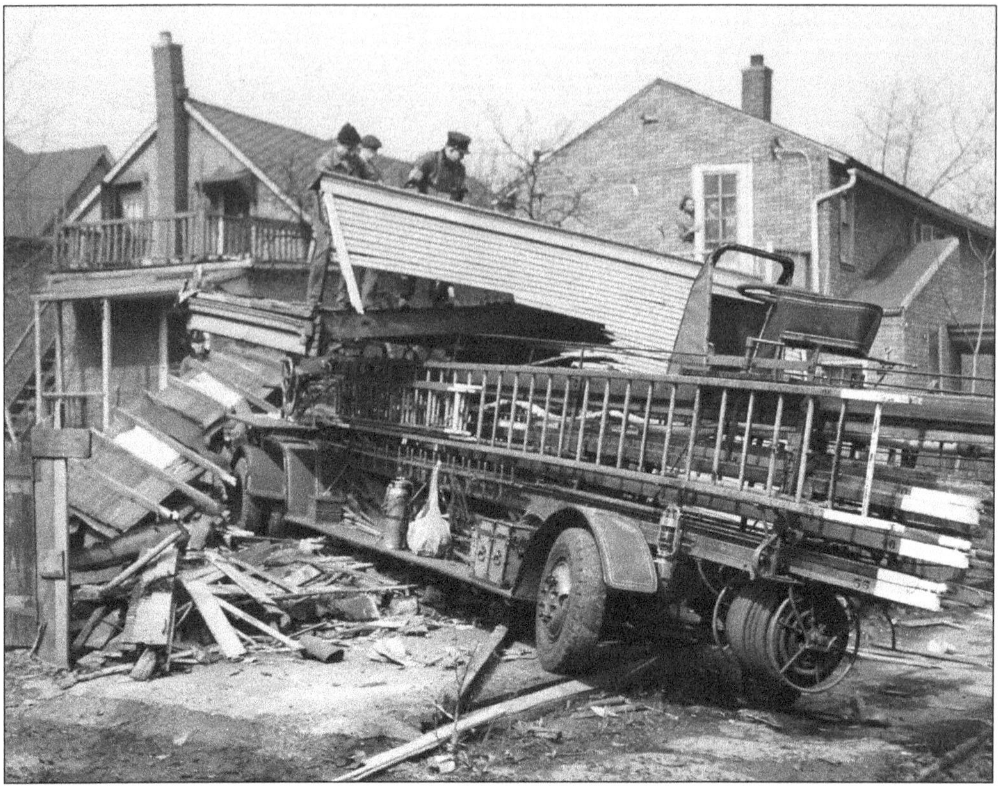

Ladder No. 6, responding to a fire, failed to negotiate a jog on Elmwood Avenue near Antietam Avenue and struck a parked car and demolished a garage. (Courtesy of Matt Lee.)

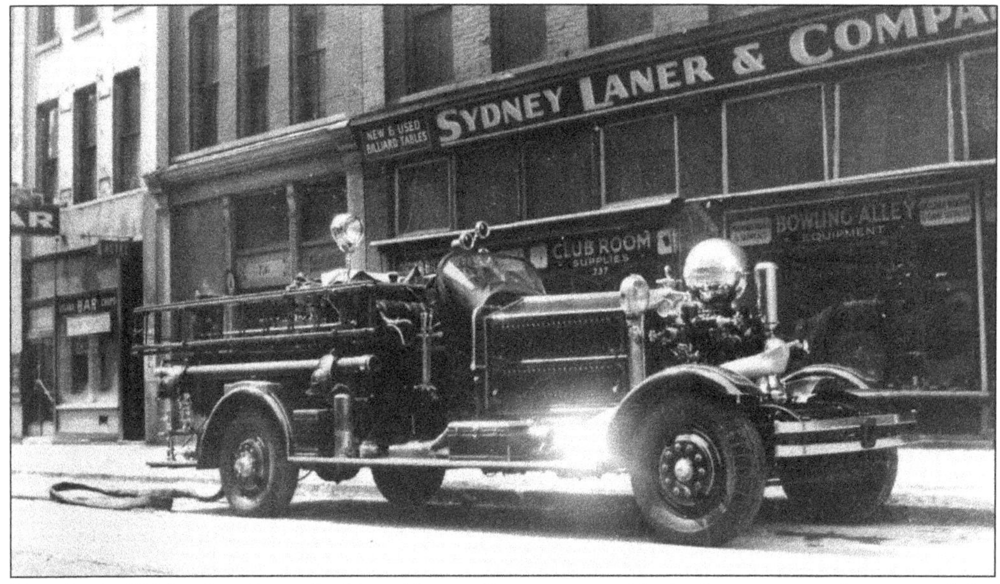

True fire buffs like this classic-looking Ahrens-Fox pumper with the front mounted pump and the chrome ball, which was for relieving air pressure. This style of vehicle, with the shinny chrome, hanging helmets, mounted lanterns, and miscellaneous tools mounted around the vehicle, was what attracted young boys and men to the fire service. (Courtesy of Matt Lee.)

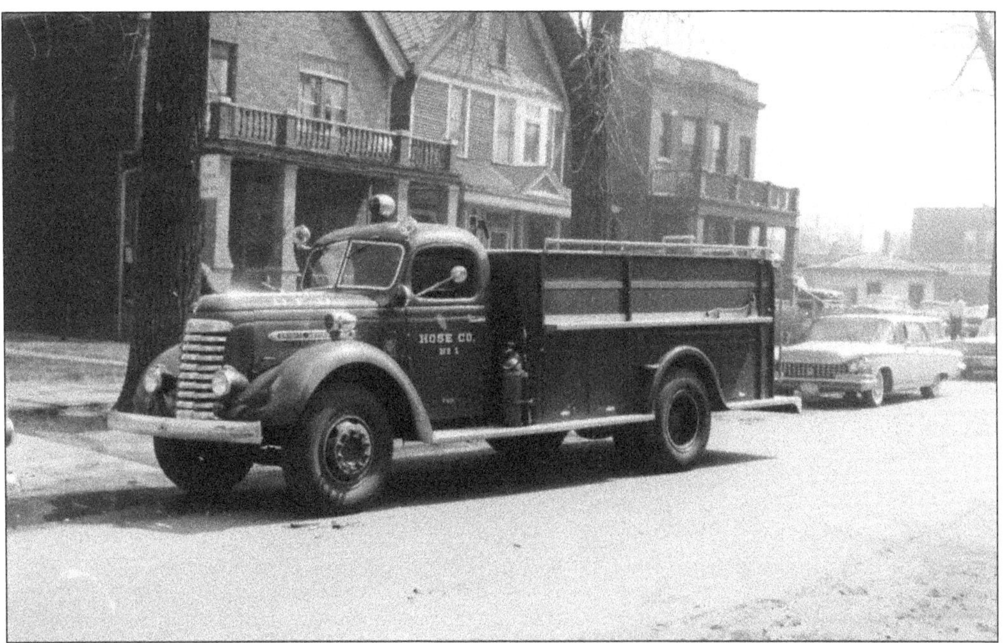

This General Motors Corporation (GMC) fire apparatus served as Hose Company No. 1. The vehicle had several thousands of feet of three-inch hose on board that responded to extra-alarm fires when needed. (Courtesy of Matt Lee.)

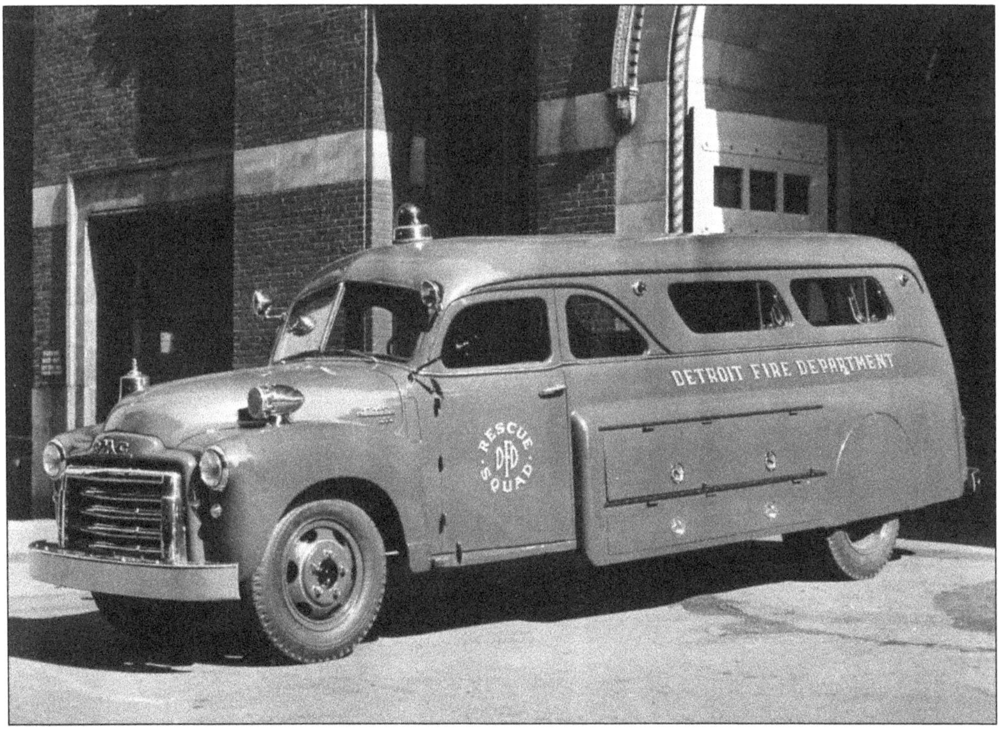

Pictured is Rescue Company No. 1 shortly after moving into fire headquarters, located on Larned Street. This 1948 squad car had its body constructed by Proctor-Keefe Company of Detroit, and it was placed on a GMC chassis. (Courtesy of Matt Lee.)

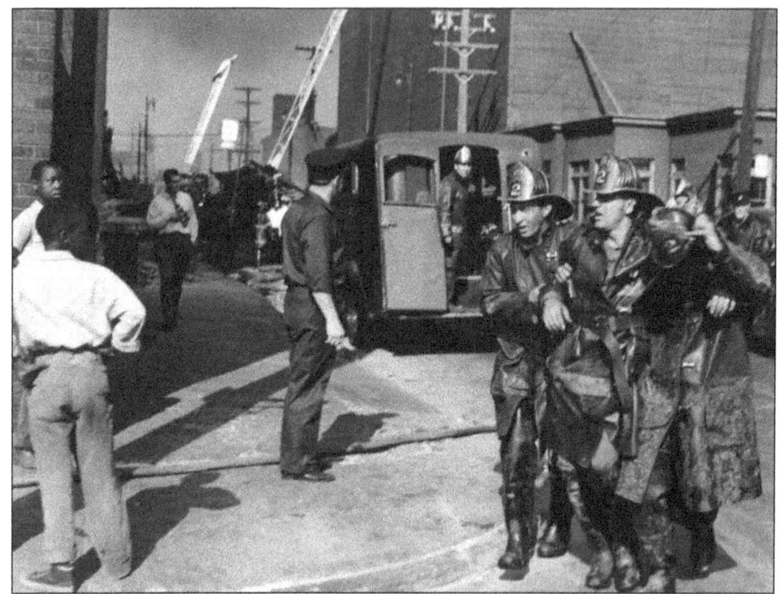

At this extra-alarm fire scene, firemen of Ladder Company No. 2 assist their fellow fireman to a safe area away from the blaze to seek medical assistance while people watch. (Courtesy of Matt Lee.)

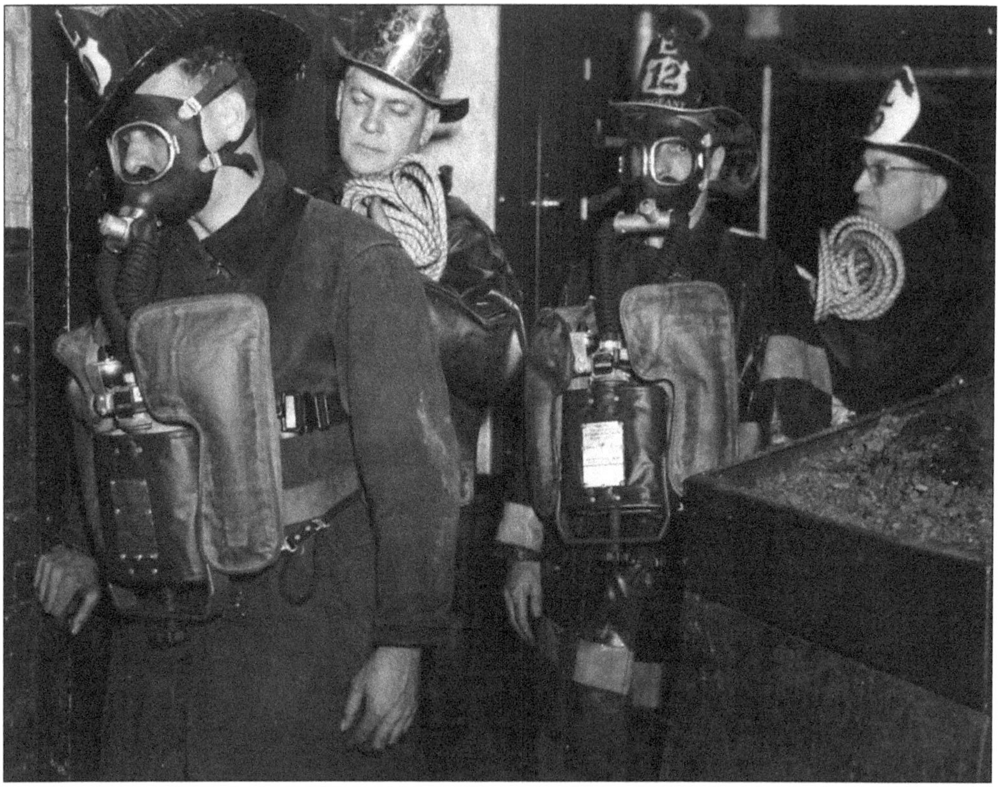

Firemen from Ladder Company No. 9 and Engine Company No. 12 are wearing the Chemox breathing mask and are preparing to enter a building on fire, while other members tie rope to each member to help retrieve them if they were unable to exit the building. (Courtesy of Matt Lee.)

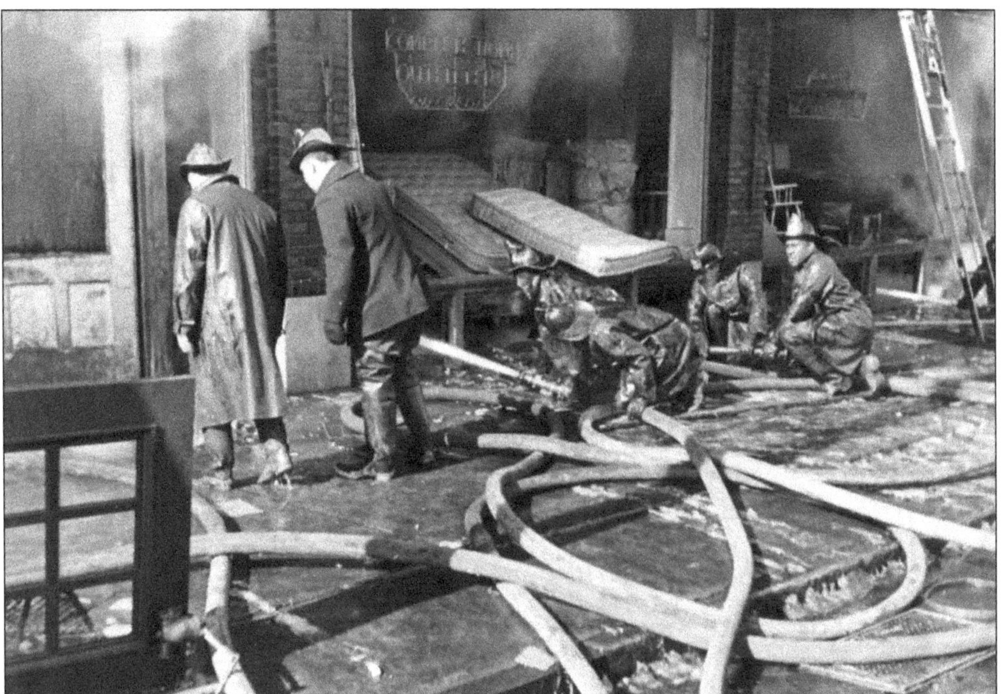

At the Ray Furniture Store fire at Michigan and Livernois Avenues, Fireman Joseph R. Schneider of Ladder Company No. 22 lost his life when he fell through the floor into the blaze below in the basement. This view of the Livernois Avenue side shows the crew of Engine Company No. 34 working hand lines into the basement. This stubborn fire took six hours to bring under control. At the far right is Fireman Marcena W. Taylor; he was the first African American fireman to become a battalion chief for the DFD. (Courtesy of Matt Lee.)

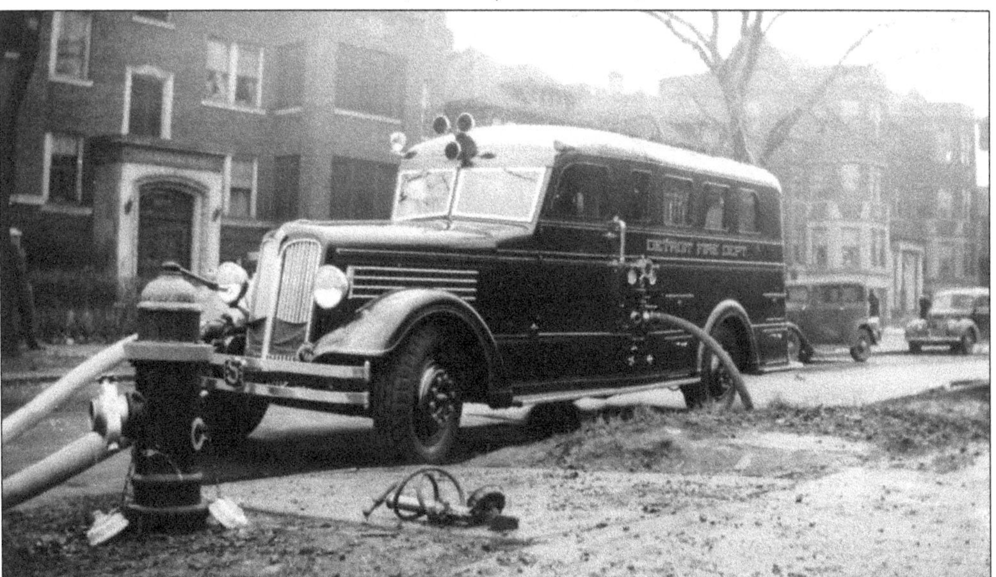

A Seagrave sedan is connected to a Detroit hydrant, blowing air out of the bottom of the base. The equipment was used to blow air into the hydrant and water out to prevent freezing in the wintertime as well as water accumulating in the hydrant's barrel. (Courtesy of Matt Lee.)

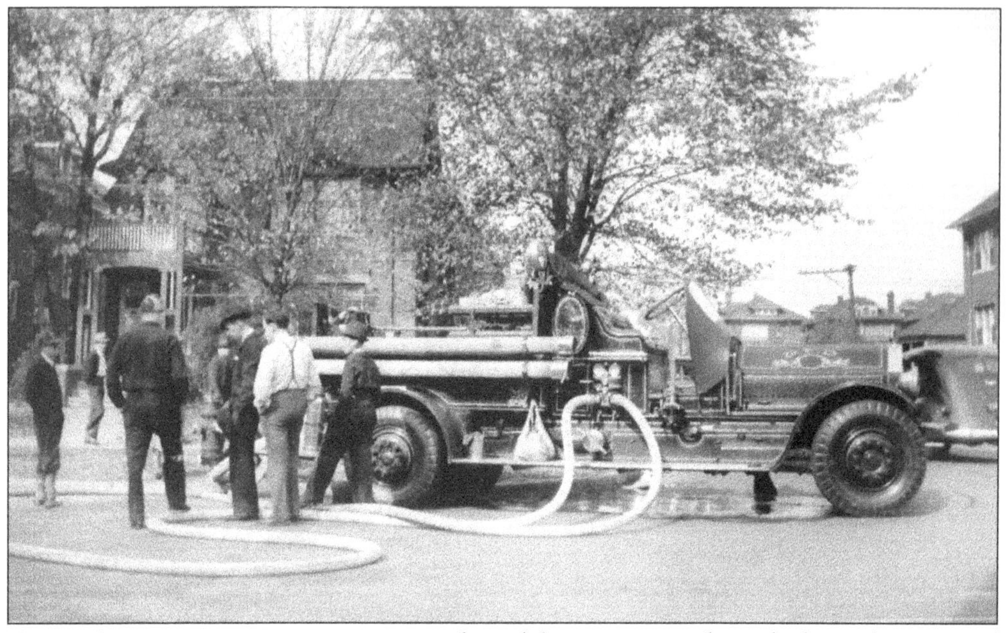

Here, a Seagrave engine pumping at a residential fire is connected to a hydrant. (Courtesy of Matt Lee.)

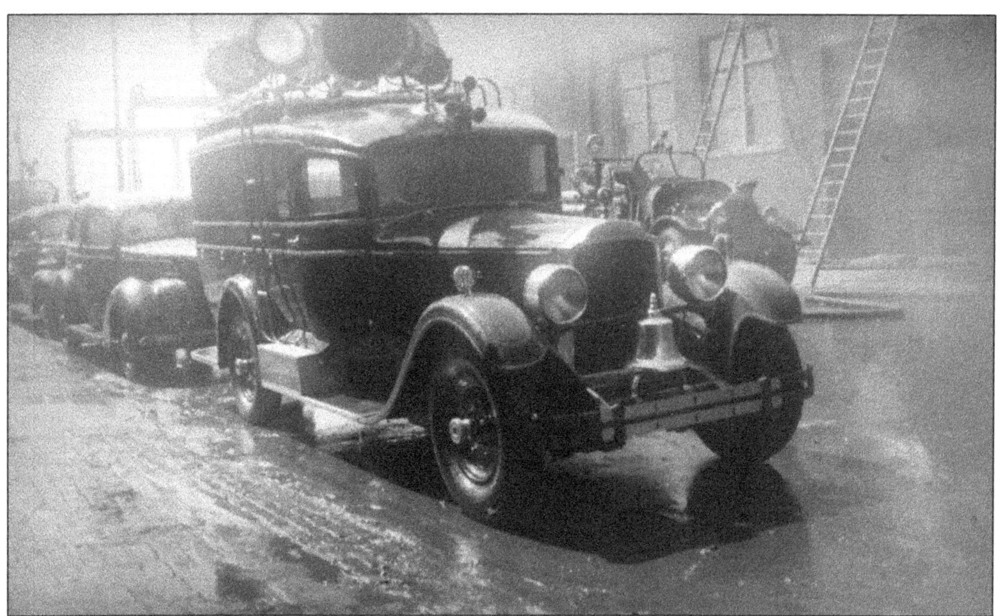

This special unit was adapted by the repair shop for use as a light wagon; the 1927 Packard was previously used as Ambulance No. 1. The light wagon was used at incidents where auxiliary lighting was needed for illumination. (Courtesy of Matt Lee.)

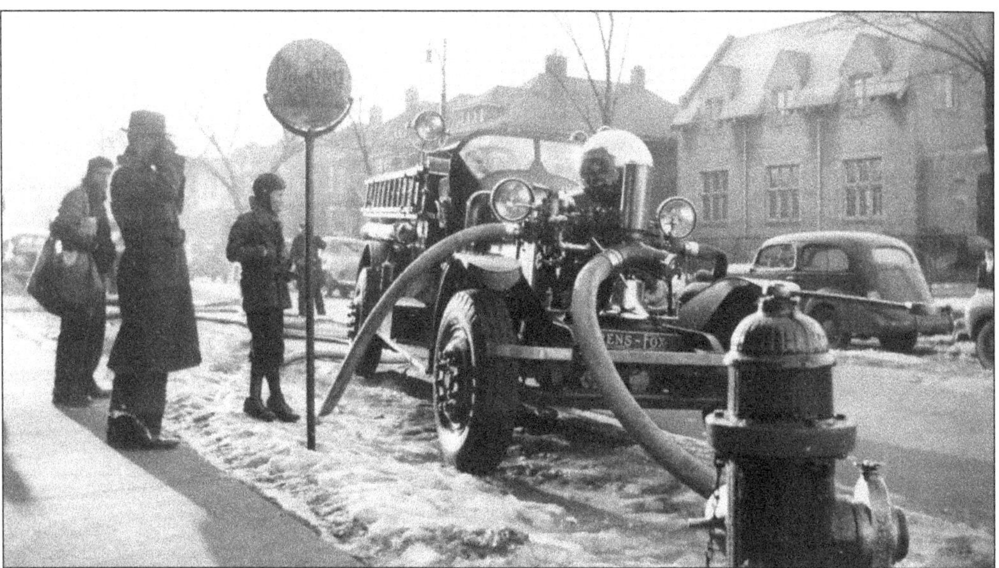

Engine Company No. 39 pumps water to two hose lines at a residential fire at Byron and Taylor Streets. This open-cab, front-mounted pumper, referred to as "the Fighting Face," was on an Ahrens-Fox. The engine was connected to a hydrant on a cold winter day while onlookers take in the scene. (Courtesy of Matt Lee.)

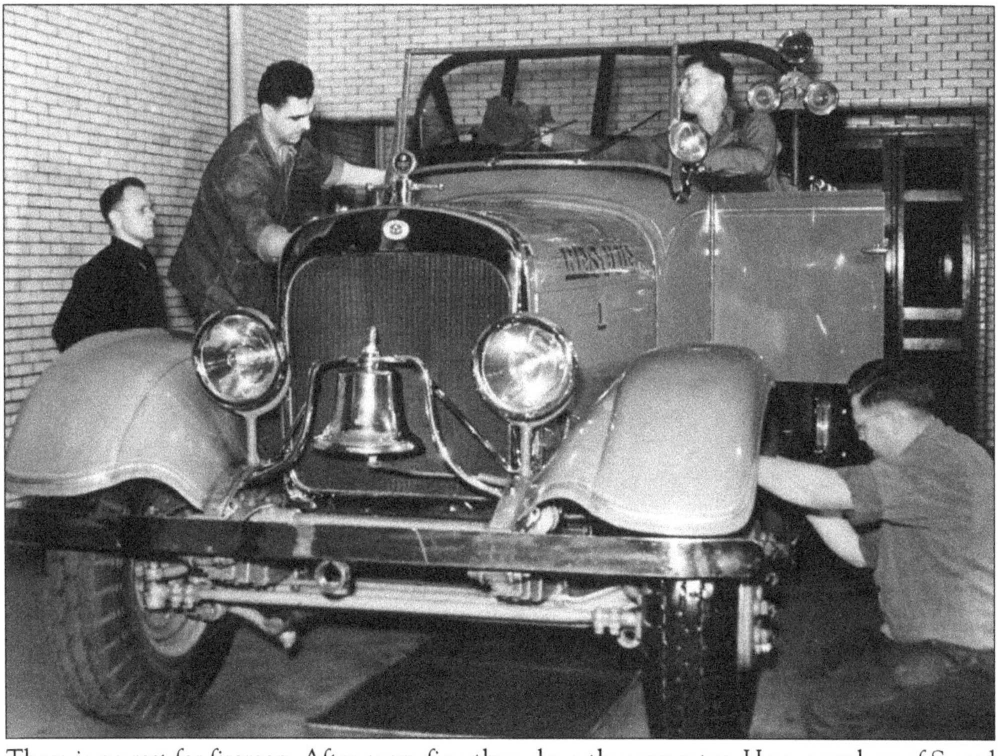

There is no rest for firemen. After every fire, they clean the apparatus. Here, members of Squad Company No. 1, located at Detroit Fire Department Headquarters, wash the vehicle to make sure it is clean. If any of the chief officers were stationed there, they would inspect the rig. Also, visitors who came the firehouse liked to see clean vehicles. (Courtesy of Dan Jasina.)

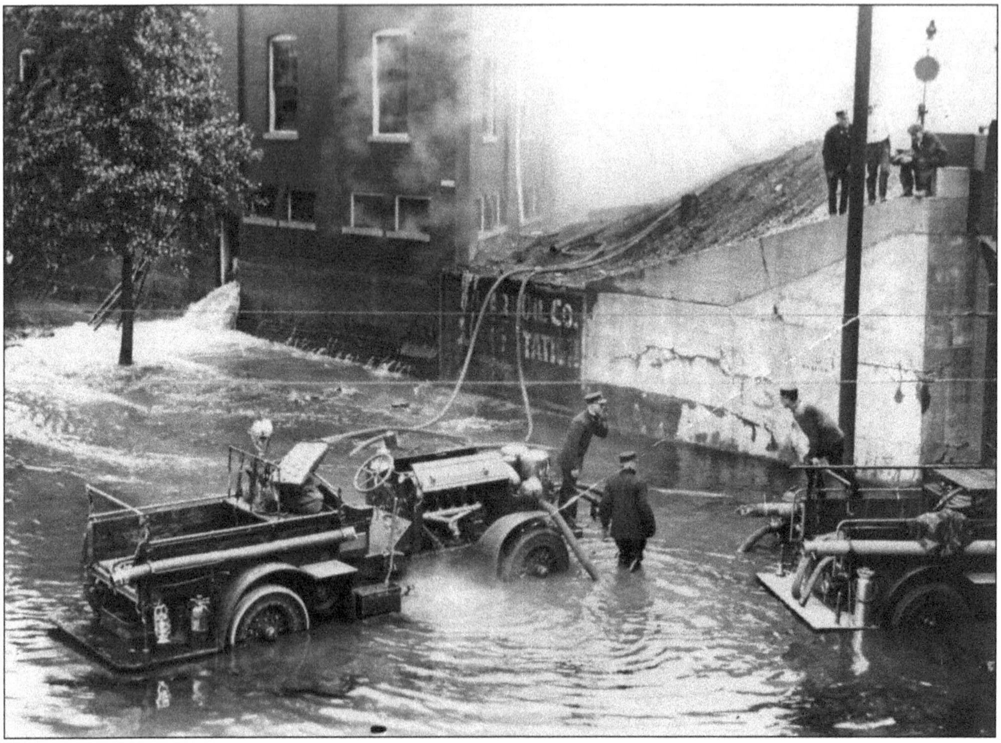

A Detroit fire engine sits submerged while pumping water to hose lines in an overflow path of water coming from a fire at the Standard Oil Company, located at Scolien Street and Michigan Avenue. Due to being in a low spot along the railroad tracks, water was starting to rise where the tail boards were becoming covered in water. (Courtesy of the Detroit Firemen's Fund.)

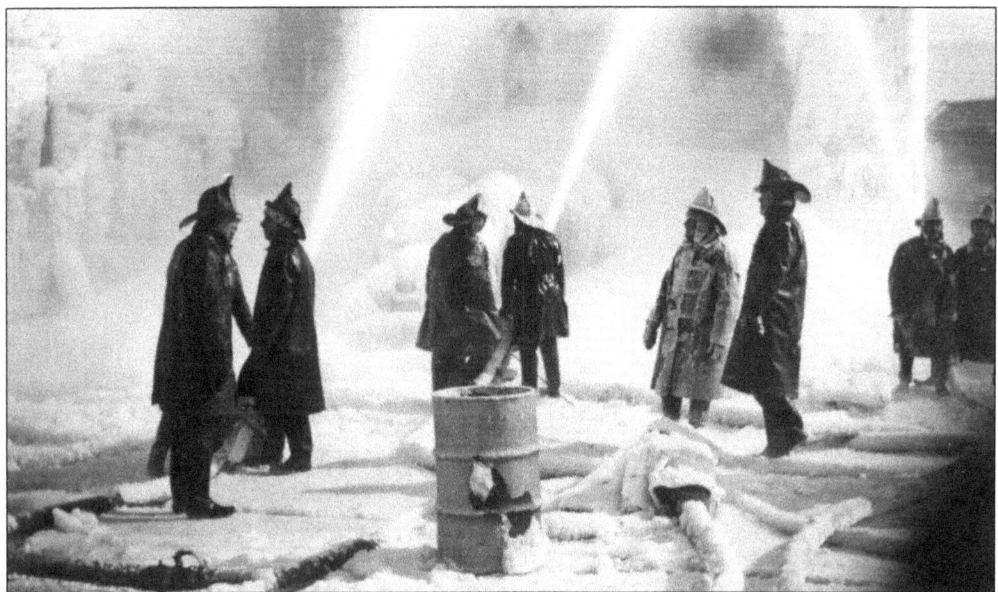

Detroit firemen are operating deck guns directed toward the building. This fire occurred in the winter, and the building, ground, and hoses are all coated in ice. The photograph is of the "Malt House fire" on Grand River Avenue around 1930. (Courtesy of the Detroit Historical Society.)

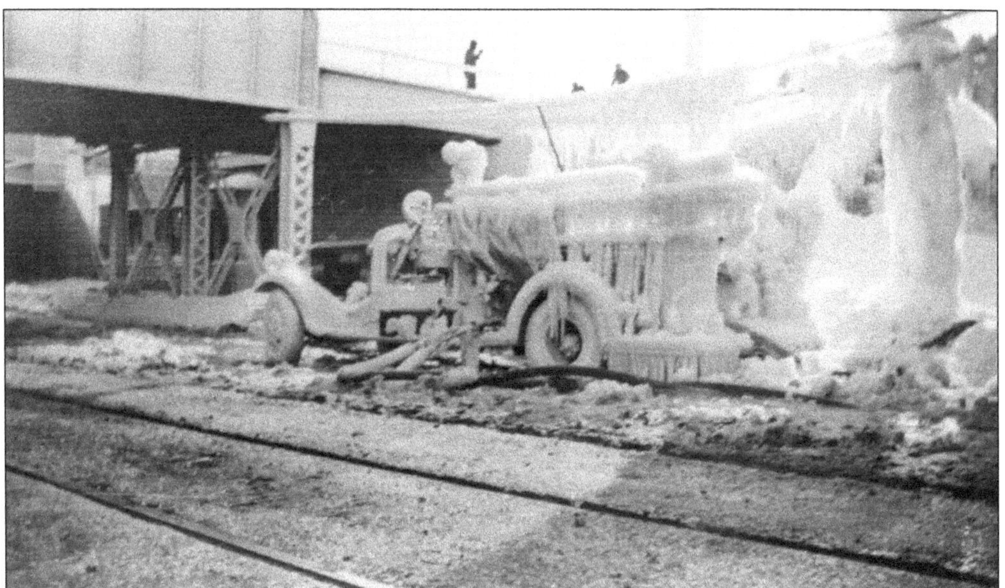

A fire department vehicle covered in ice is parked along an ice-covered building that can be seen in the right background. Two overpasses can be seen in the left background. Railroad tracks are visible in the foreground. This photograph is from the Malt House fire on Grand River Avenue around 1930. (Courtesy of the Detroit Historical Society.)

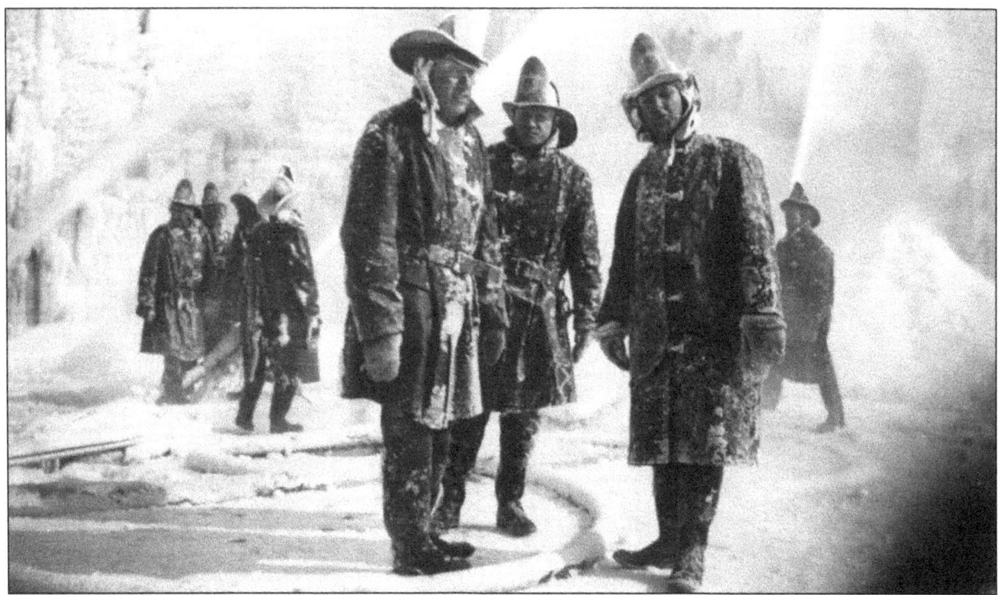

Detroit firemen from Engine Company No. 5 fight a fire at a building covered in ice and snow. In the background, other firemen are operating fire hoses and deck guns directed toward the ice-covered building. This photograph is from the Malt House fire on Grand River Avenue around 1930. (Courtesy of the Detroit Historical Society.)

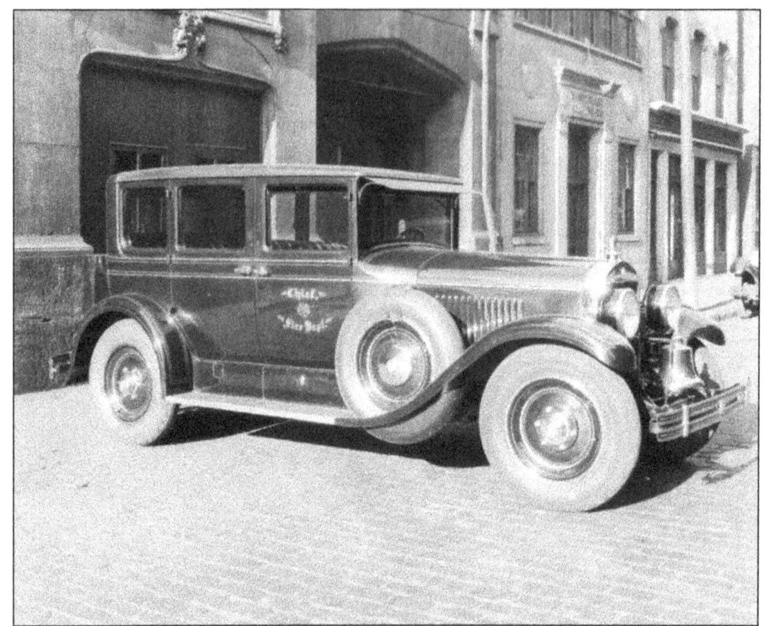

The chief's car is seen here in front of the Fire Department Headquarters on West Larned Street in 1927. (Courtesy of the Detroit Historical Society.)

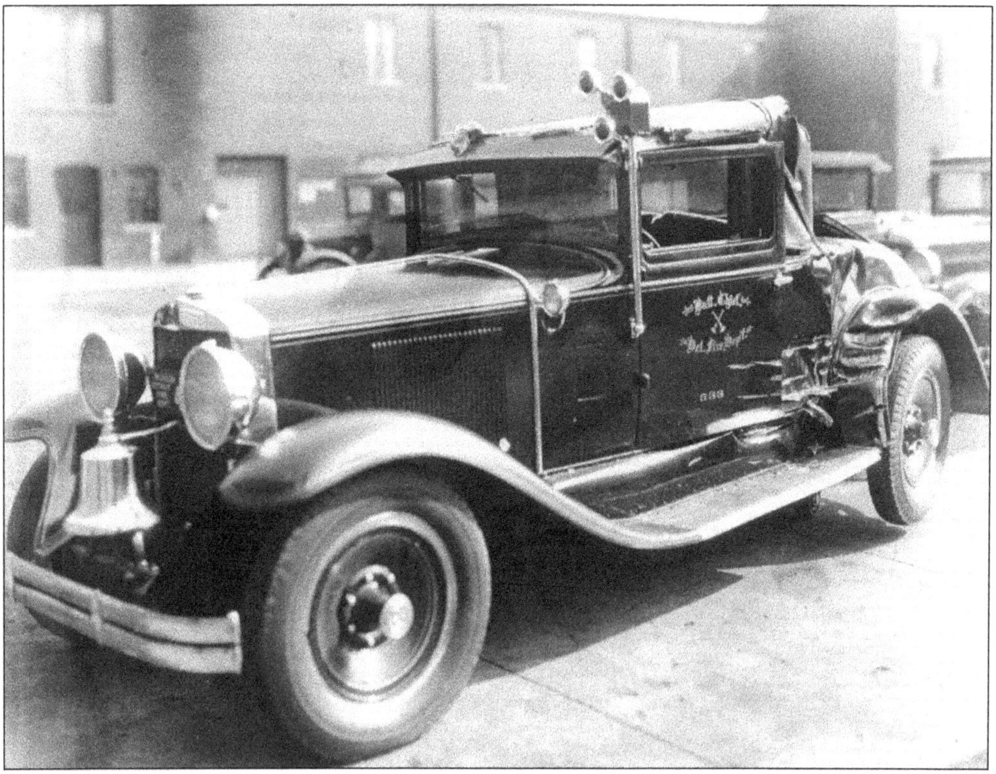

A battalion chief's vehicle, which was involved in an accident, is visibly damaged on the driver's-side rear section. (Courtesy of the Detroit Firemen's Fund.)

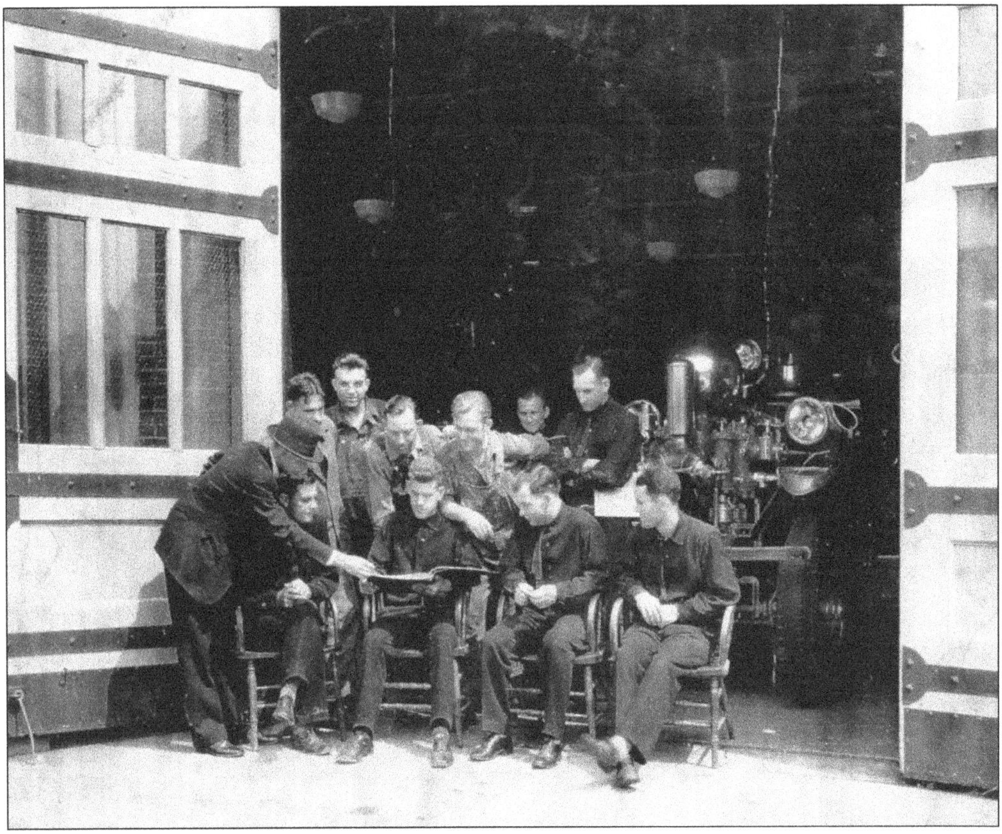

Firemen from Engine Company No. 17 are pictured looking at a book or album and seem quite interested in its contents. Behind them is Engine Company No. 17's Ahrens-Fox. The swing-out apparatus doors that Detroit Fire Department was known for on its firehouses are open so members could be outside in the sunlight. (Courtesy of the Detroit Firemen's Fund.)

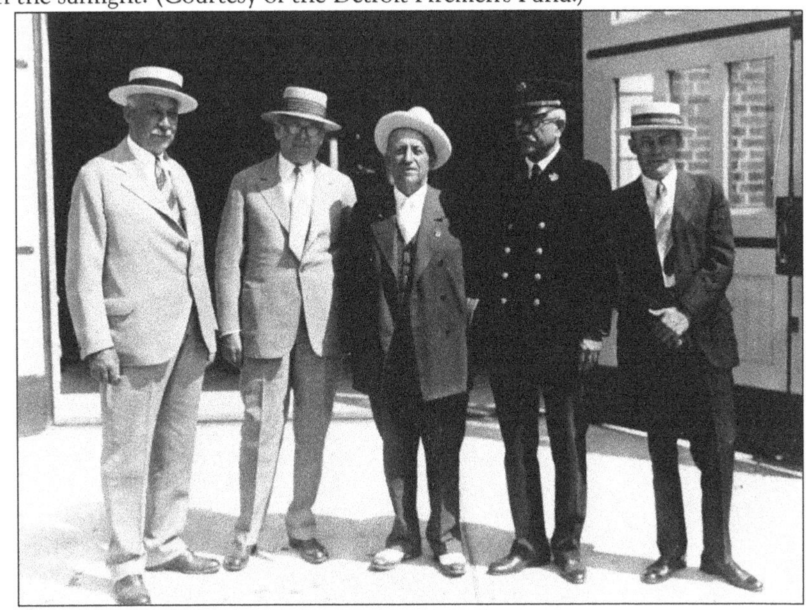

Detroit Fire Department chief Stephan DeMay poses outside Engine Company No. 17's firehouse with members of the board of commissioners. (Courtesy of the Detroit Firemen's Fund.)

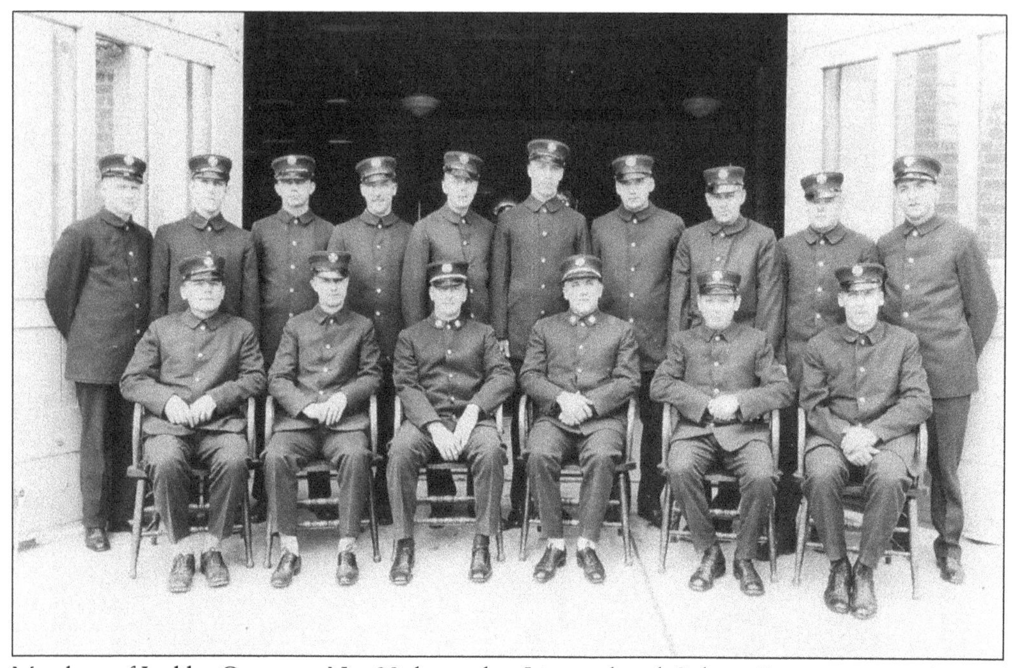

Members of Ladder Company No. 28, located at Linwood and Calvert Streets, pose in front of their firehouse shortly after being placed in service. Ladder Company No. 28 was established in 1927. (Courtesy of the Detroit Firemen's Fund.)

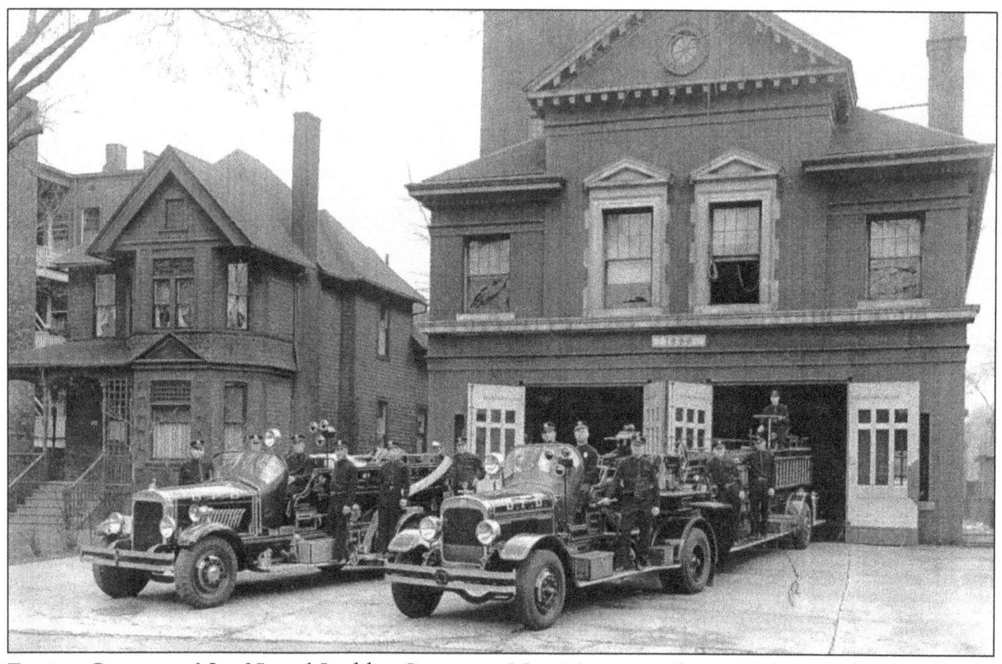

Engine Company No. 35 and Ladder Company No. 15 pose in front of their firehouse shortly after Engine No. 35 moved in 1940. This firehouse opened up with Engine Company No. 24 in service, but the company was taken out of service and disbanded in 1940. Engine Company No. 35 was established in 1911 and came from its firehouse at Mount Vernon and Beaubien Streets. (Courtesy of the Detroit Firemen's Fund.)

Four

1950s–1960s

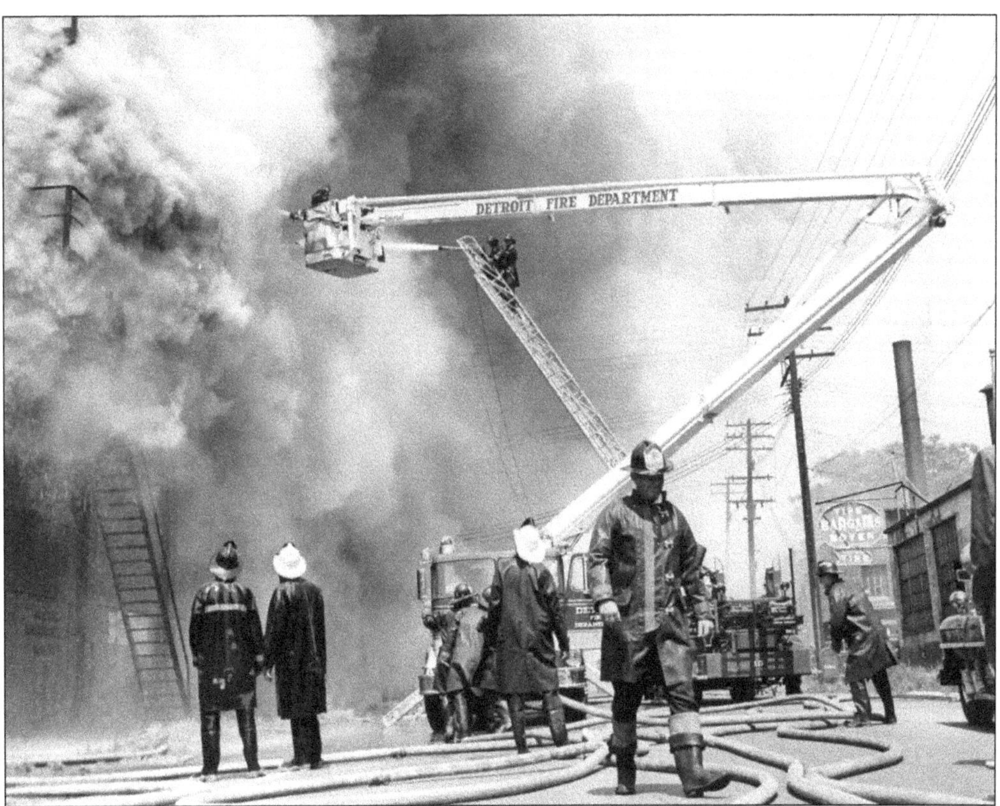

The DFD Snorkel Company No. 1 operates at the scene of a five-alarm fire in this large, five-story warehouse on Detroit's near east side. It was the second five-alarm fire in the last 24 hours and resulted in two firefighters being injured and Ladder Company No. 10 being damaged by a falling wall. (Courtesy of Matt Lee.)

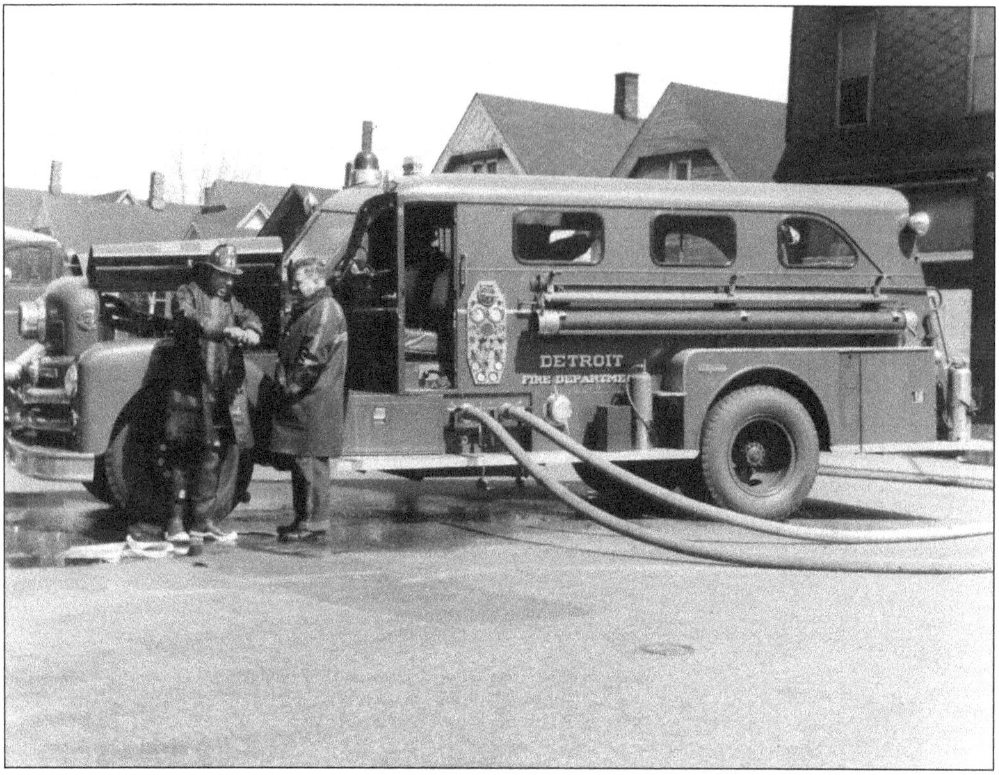

Here, Engine Company No. 23 is working at a residential fire on Detroit's east side. A common issue with the Seagrave sedan was the engine compartment always had to have the engine doors raised to prevent overheating. A fireman is pictured ringing out a pair of wet gloves while talking to the FEO (fire engine operator). (Courtesy of Matt Lee.)

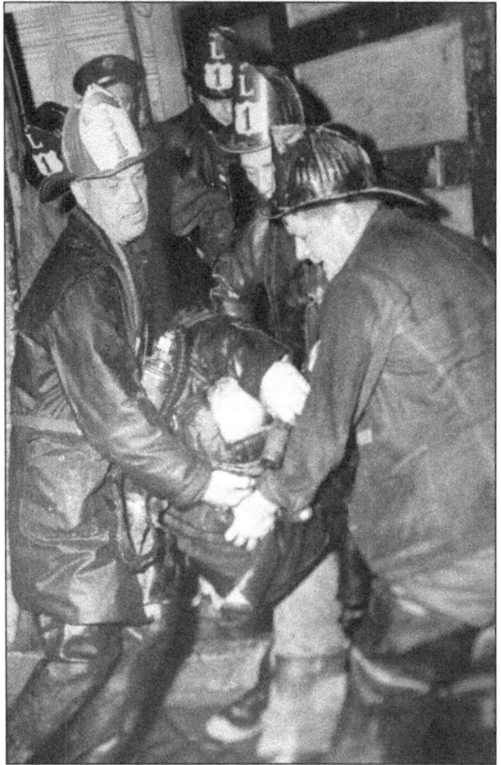

Members of Squad Company No. 1 and Ladder Company No. 1 help drag an unconscious fireman out of a building. He collapsed several rooms deep in this apartment building and had to be carried out. (Courtesy of Bill Eisner.)

Detroit firemen have always been known for taking care of their tools and equipment. Here, two firemen are polishing the brass nozzles and hose fittings from their engine as ordered by the captain of the company. (Courtesy of Dan Jasina.)

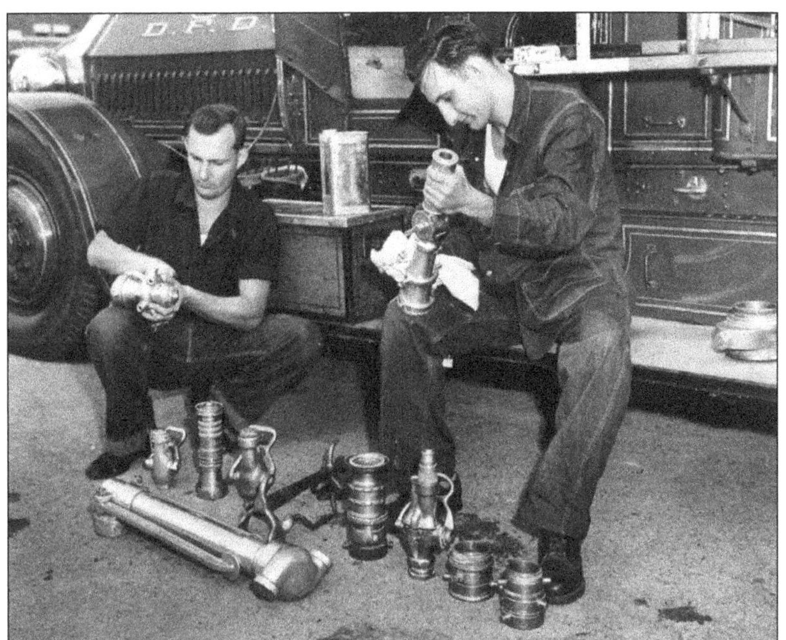

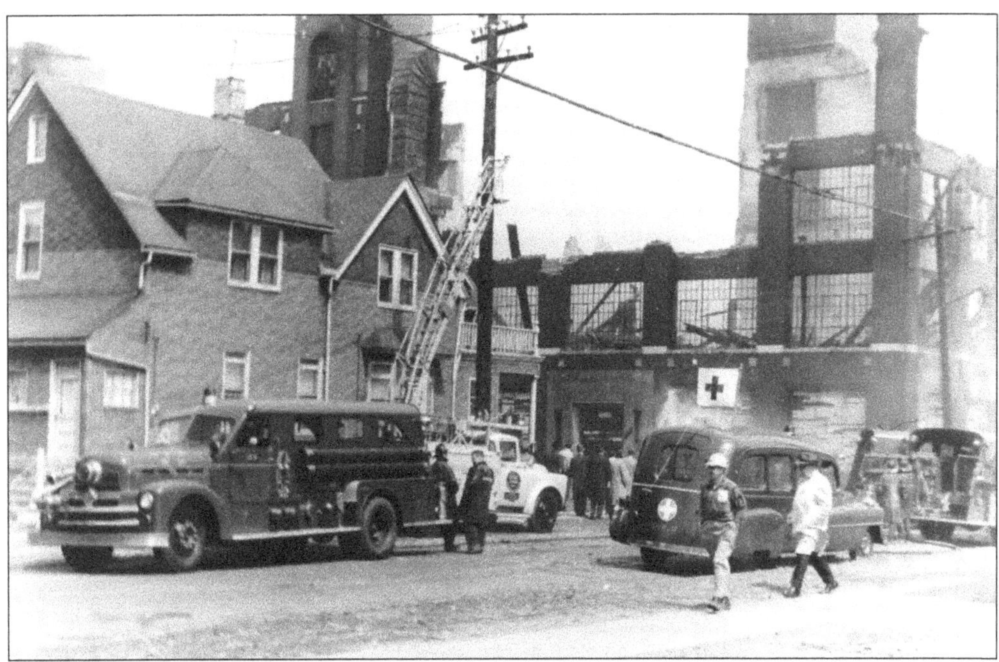

At an extra-alarm fire, several fire companies continue to hit hot spots, while other companies perform overhaul. Engine Company No. 23 and Ambulance No. 1 work the blaze while waiting to be released from the scene to be able to return to their firehouse. This is another example of the DFD performing medical service in that era. (Courtesy of Matt Lee.)

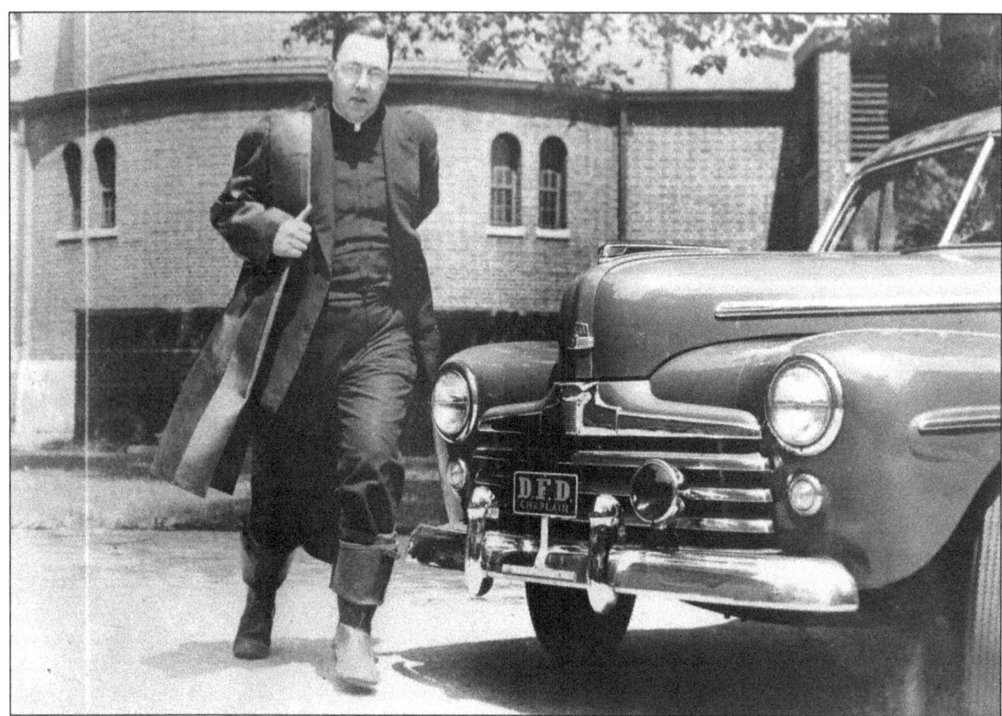

The Detroit Fire Department's chaplain, Fr. John O'Connor of St. Benedict Catholic Church in Highland Park, is seen putting on his turnout coat while responding to a multiple-alarm fire in his Ford sedan. Note the license plate and warning lights located in the front grill. (Courtesy of Matt Lee.)

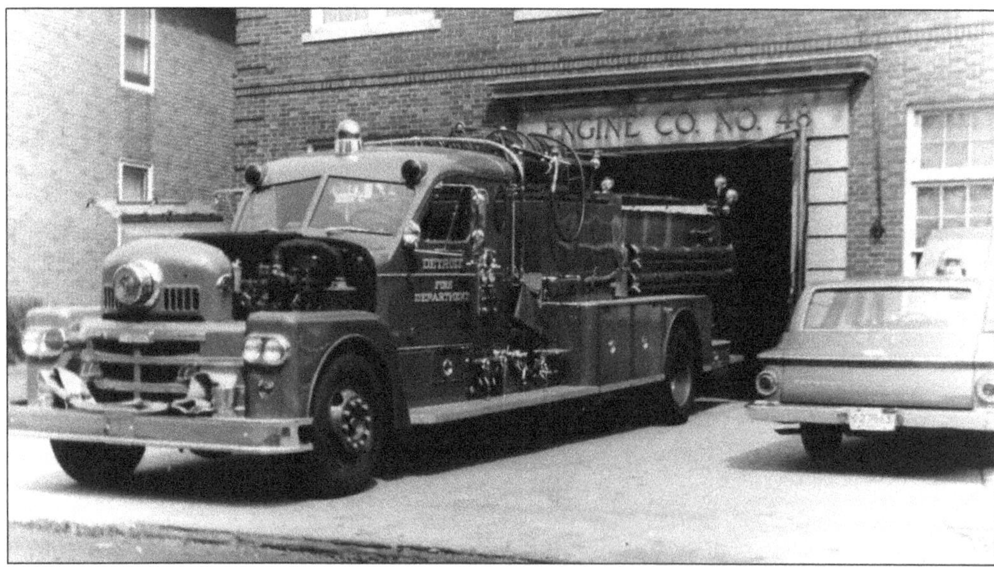

Engine Company No. 48 had a Seagrave "QUAD," the only one in the city of Detroit due to its first-response area having limited access, meaning Engine Company No. 48 could be the only company at a fire for some time because of limited access. This vehicle had a hose bed, pump, water tank, and an assortment of portable ground ladders. The high-pressure booster lines are located on top of the vehicle. (Courtesy of Matt Lee.)

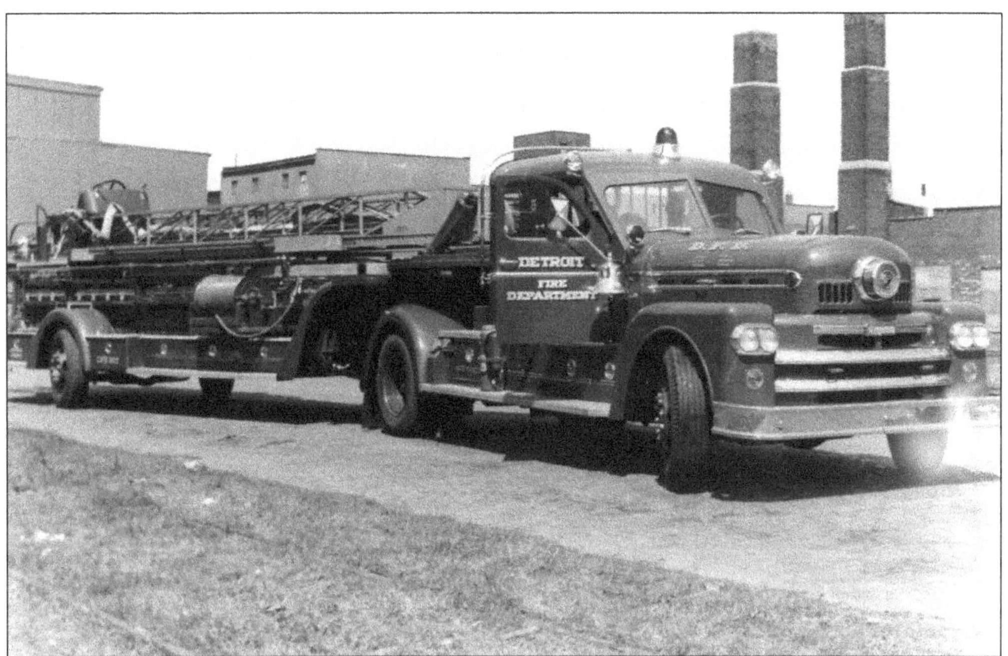

Ladder Company No. 5's Seagrave 100-foot midship aerial ladder was on a two-piece tractor. Shown on the officers' side of the vehicle is the red line that supplied an onboard water supply, which was normally found on ladder trucks. The aerial ladder has a length of two-and-half-inch hose preconnected to the aerial pipe alongside the main ladder. (Courtesy of Matt Lee.)

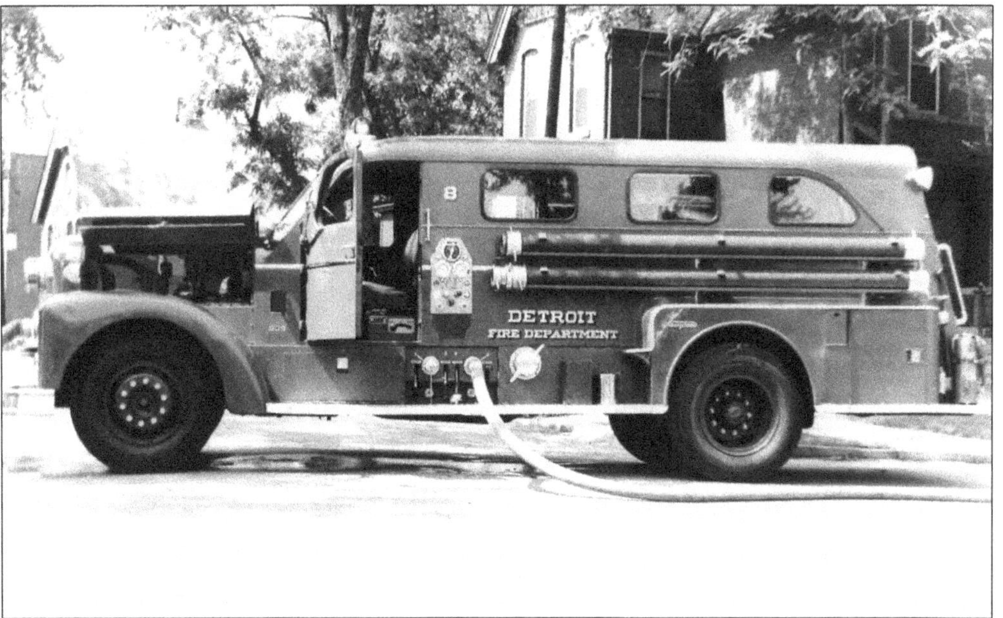

Engine Company No. 8 was operating here at a residential box alarm in central Detroit. Again, with the Seagrave sedan the famous lifting of the engine hood doors took place to allow heat to escape to prevent overheating. The view of the driver's side shows the sleek but compact pump panel that is all in one area, which was user friendly. The hard-suction hose was mounted so the FEO could lift the hose off the vehicle and hook up to the hydrant. (Courtesy of Matt Lee.)

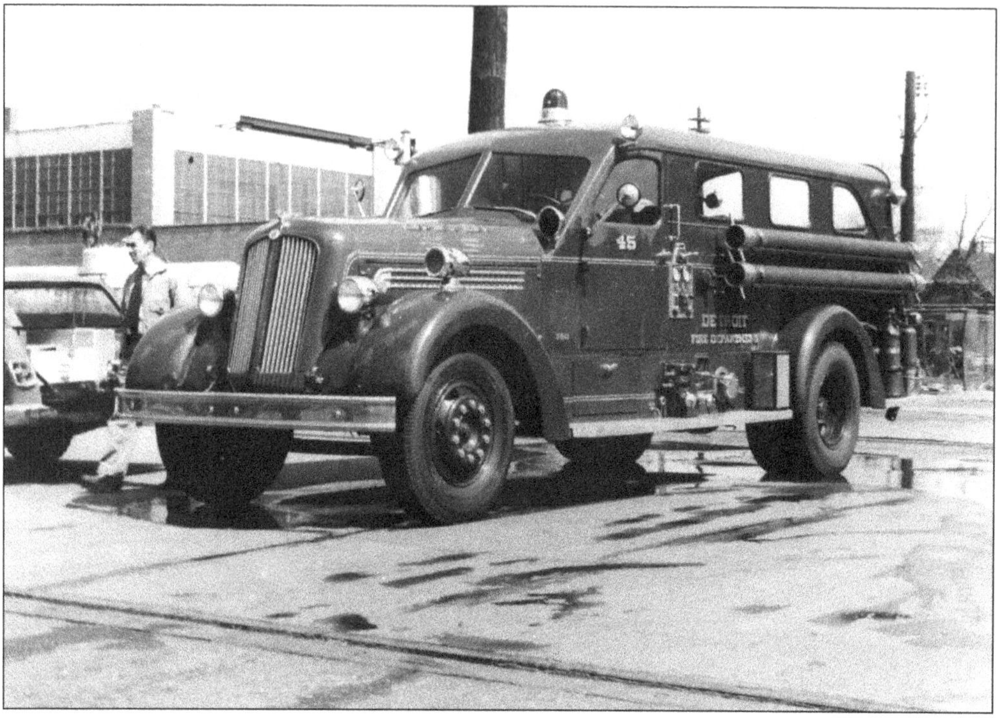

Engine Company No. 45 had just finished working at this fire. Seagrave sedans were normally seen at fires in Detroit from the 1930s through 1970s. (Courtesy of Matt Lee.)

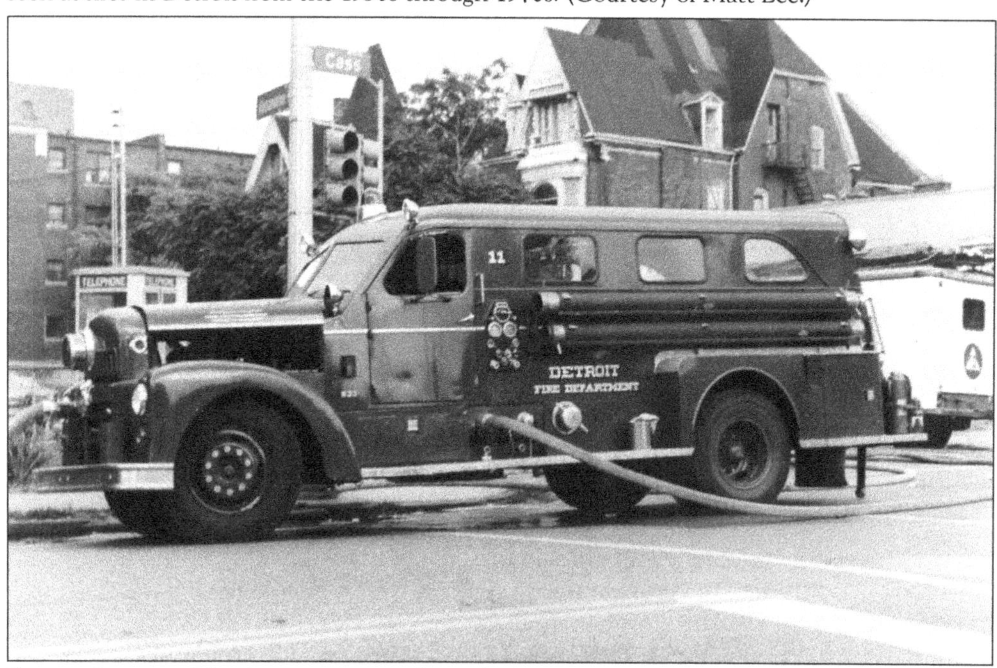

Engine Company No. 11 was connected to a hydrant supplying a hose line at this commercial building fire, located at Cass Avenue and Alexandrine Street. The famous-to-Detroit-style vehicle, the Seagrave sedan had such features like the compact pump panel and the two pieces of hard-suction hose mounted to the driver's side of the vehicle. (Courtesy of Matt Lee.)

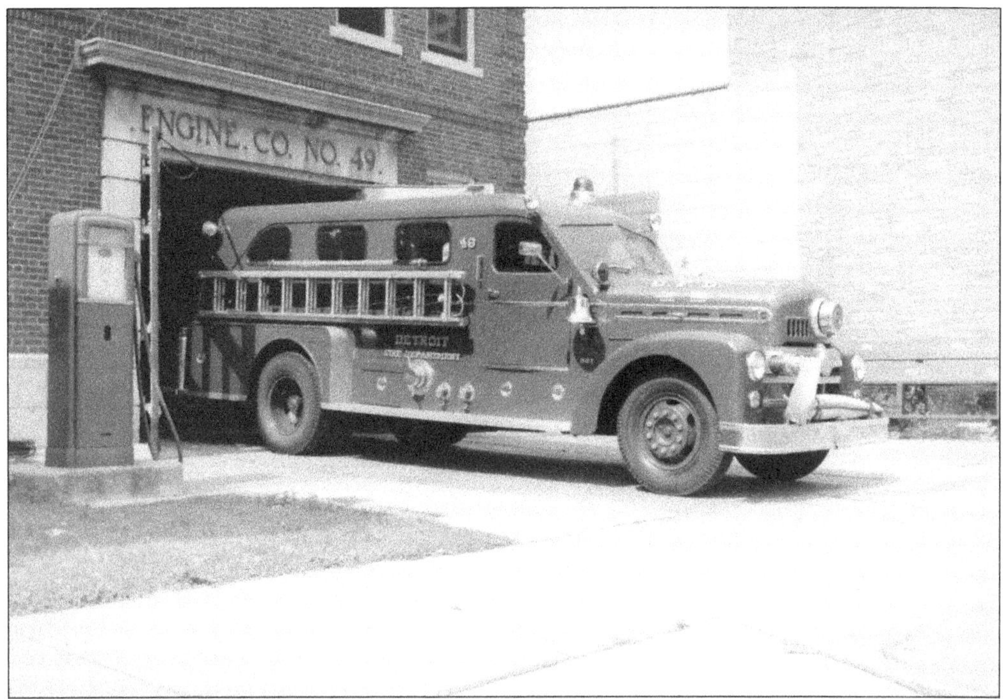

Engine Company No. 49 is parked in front of its firehouse, located at Grand River Avenue and Manor Street on Detroit's west side. The famous-to-Detroit-style vehicle, the Seagrave sedan had the front hydrant connection and side-mounted portable ground ladders on the officers' side of the vehicle. The hose bed and firemen were located in the rear under an enclosed roof. Back then, some firehouses had fuel pumps on the grounds of the property. (Courtesy of Matt Lee.)

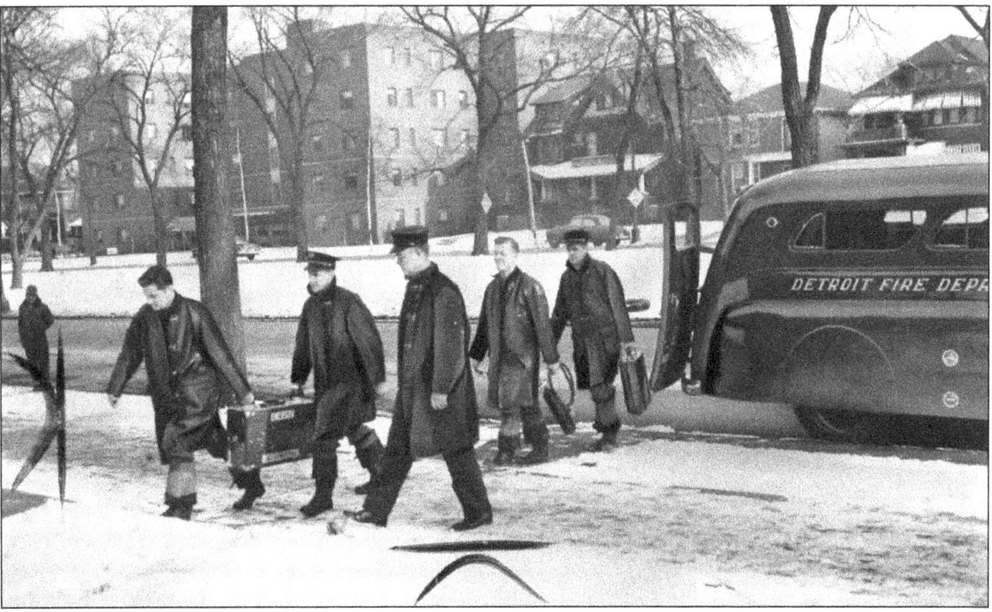

Members of Squad No. 4 responded to the scene of a medical call. Firemen have exited the rescue squad vehicle and are carrying their medical equipment into the house. (Courtesy of David Traiforos.)

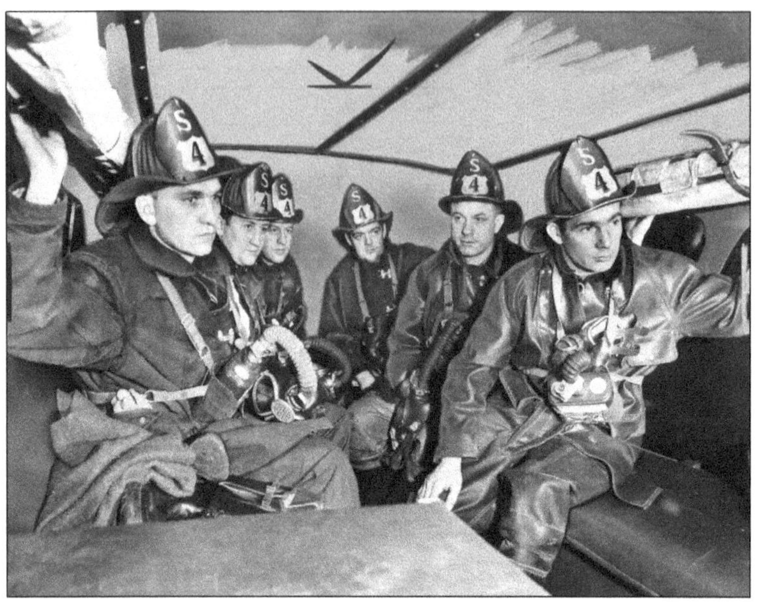

Here, members of Squad No. 4 are seated in the back of their squad vehicle preparing for arrival at the scene of a working fire. They will either man the second hose line, perform search and rescue of victims, or start looking for hidden fire that may have spread. (Courtesy of David Traiforos.)

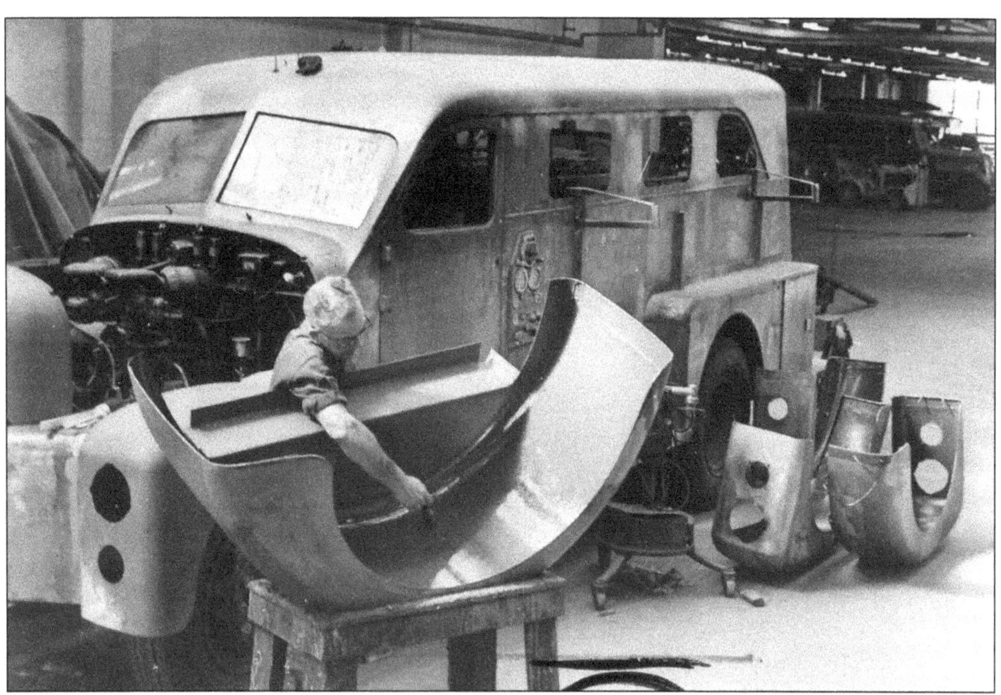

The Detroit Fire Department's repair shops stayed open 24 hours a day. With the number of fires, the vehicles were in need of constant maintenance. Here, a repairman works on a Seagrave sedan, replacing the rusted parts to make the vehicle like new so it could be put back in service. (Courtesy of Dan Jasina.)

This DFD fireman tries to get out of the wind while working at this multiple-alarm fire on a subzero winter day. The fireman's turnout gear had a collar that provided limited warmth from cold weather conditions. (Courtesy of Matt Lee.)

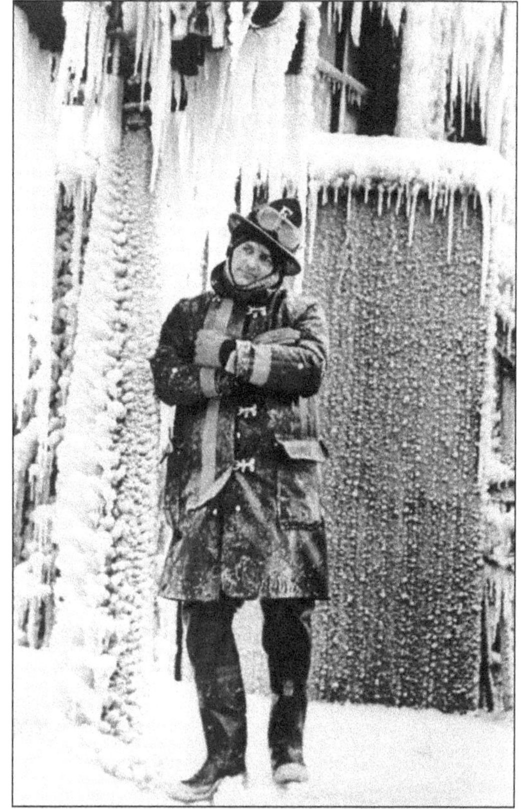

Members of the Detroit Fire Department Band pose for a photograph. The band is still performing today but with a smaller membership, which includes musicians from other fire departments. The band plays at funerals and memorial ceremonies. (Courtesy of Matt Lee.)

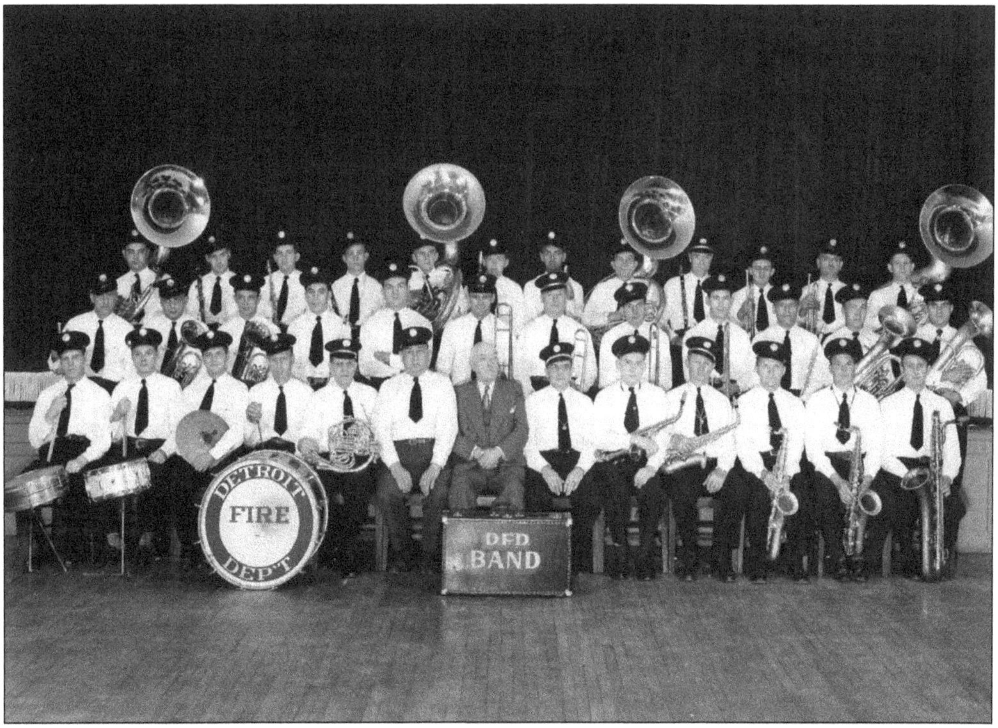

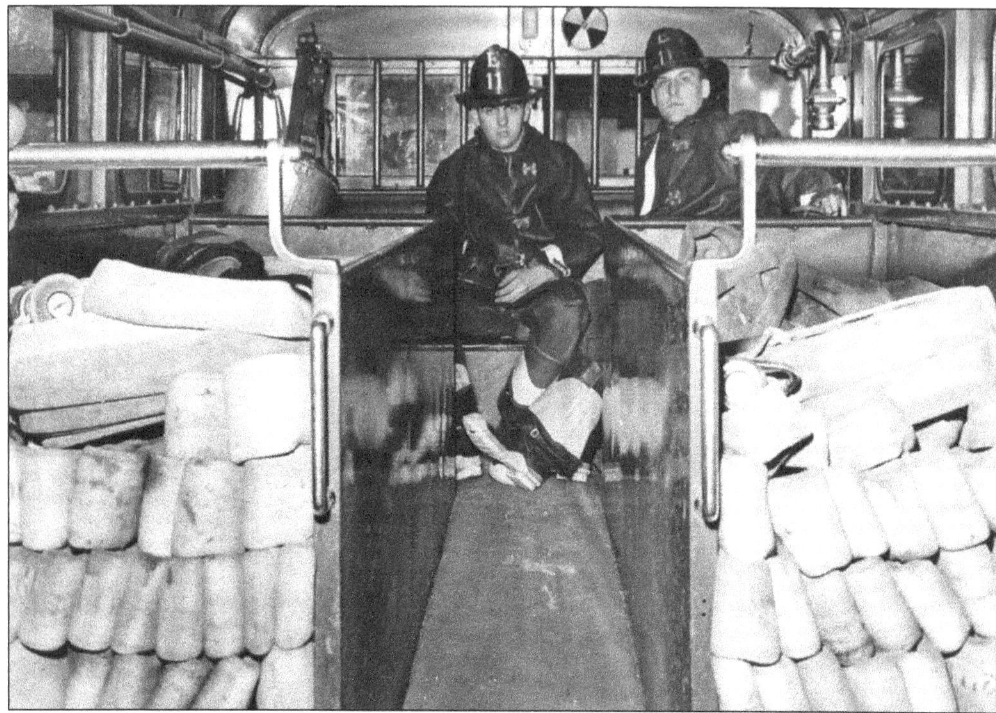

In January 1975, Engine Company No. 11 was located at Gratiot Avenue and Grandy Street prior to going out of service. Pipemen Robert Decausin (left) and John Chakin pose in the back of a Seagrave sedan; firemen were positioned along the rear hose beds and among other tools and equipment in the vehicle. (Courtesy of Matt Lee.)

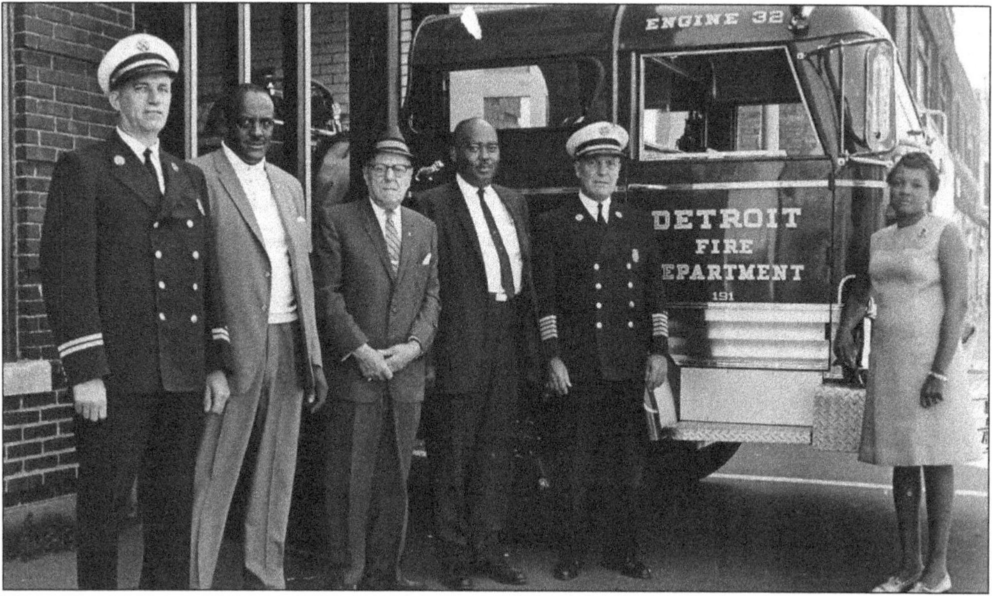

Engine Company No. 32, located at Lycaste Street and East Jefferson Avenue, received a new Mack cab forward engine in August 1967. Shown in the photograph are DFD personnel and local block club representatives. The DFD always worked with neighborhood organizations to keep them apprised of safety concerns. (Courtesy of Matt Lee.)

A fireman is on house watch at Engine Company No. 20. The watch desk was always located near the front door on the apparatus floor. A fireman was always required to be alert and on watch in case someone came to the firehouse to report a fire. The area consisted of the telegraph equipment, box response cards, telephones, journal books, and an old, wooden chair. (Courtesy of Matt Lee.)

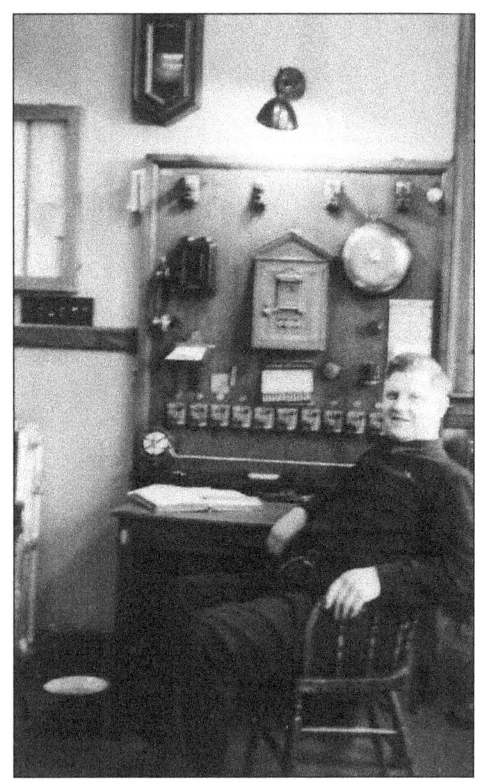

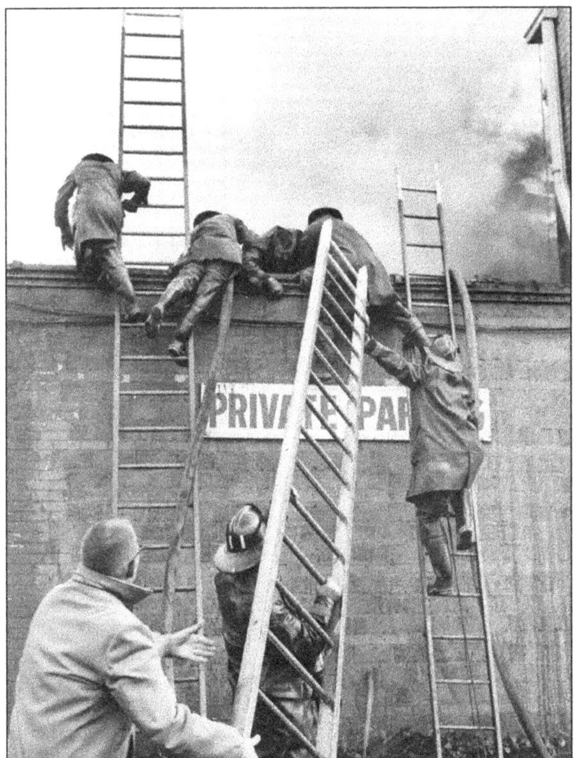

Pictured in October 1968 is a dramatic rescue of Detroit fireman John Lisuk, a member of Ladder Company No. 8, at a five-alarm fire at Vernor Highway and Livernois Avenue in a bowling alley. Fireman Lisuk started to fall through the weakened roof, holding on the parapet wall, while firemen raised wooden portable ladders to get close to him and save him from serious injury or death when the roof caved in on the second floor. (Courtesy of Matt Lee.)

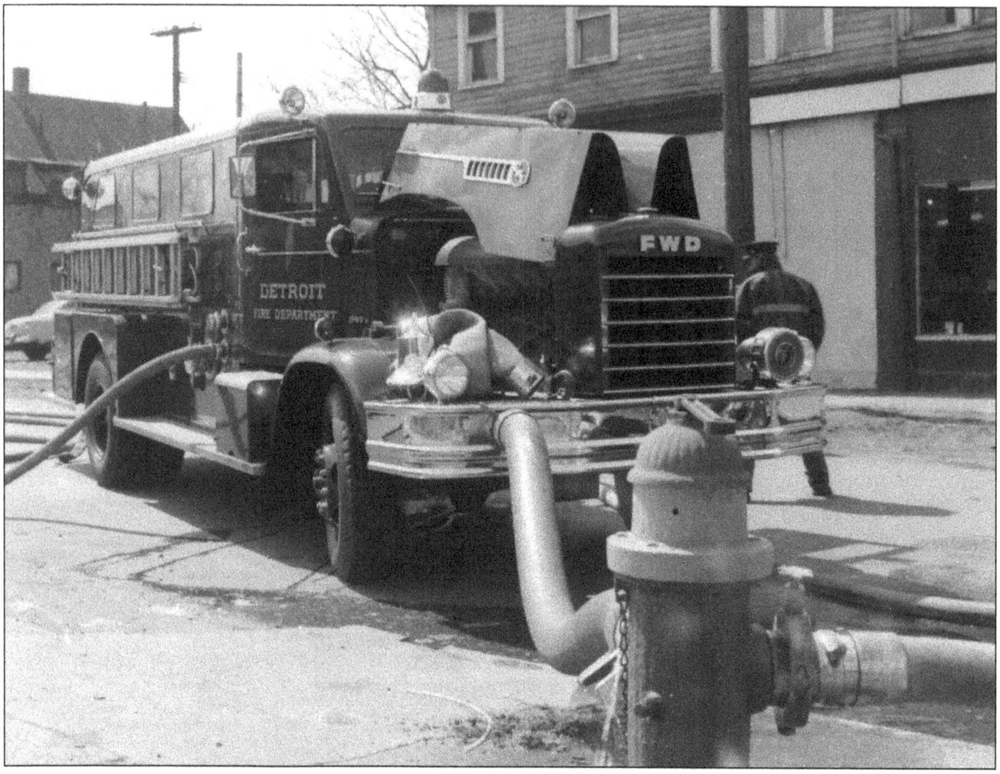

Engine Company No. 6's FWD (four-wheel drive) sedan pumper was used in Detroit to replace some of the Seagraves. Here, it is connected to a hydrant using the hard-suction hose with the engine compartment doors open to prevent overheating. (Courtesy of Matt Lee.)

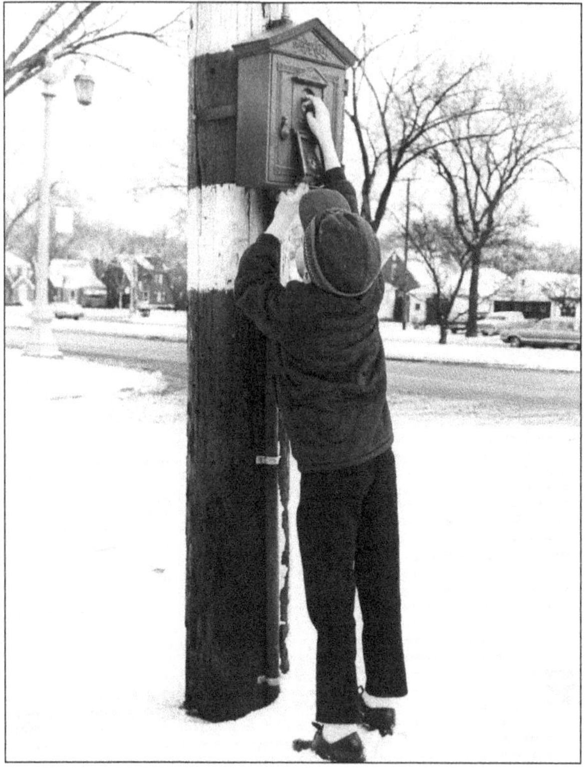

A young boy demonstrates how to pull a street corner fire-alarm box, which would summon the DFD to respond for the report of a fire or a false alarm. When the lever on the street box is pulled, it sends the fire box number to the Central Fire Alarm Office on the Gamewell system. The operator at the office references the box number and location and sends the closest fire companies to investigate why the box was activated. (Courtesy of Matt Lee.)

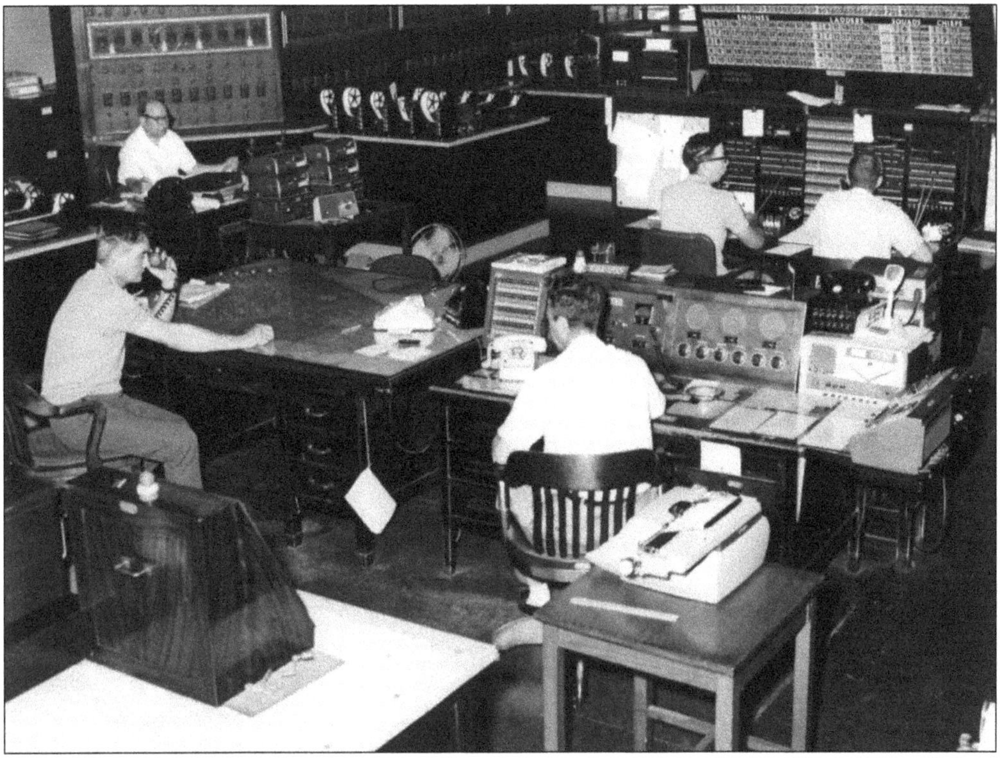

This 1965 photograph shows the second-floor nerve center of the Central Fire Alarm Office. The use of the two-way radio system was first introduced in 1929 and completely equipped in all apparatuses in 1952. Today, the Central Fire Alarm Office is state of the art and able to continue operations self-sufficiently in the event of a terrorist attack. (Courtesy of Matt Lee.)

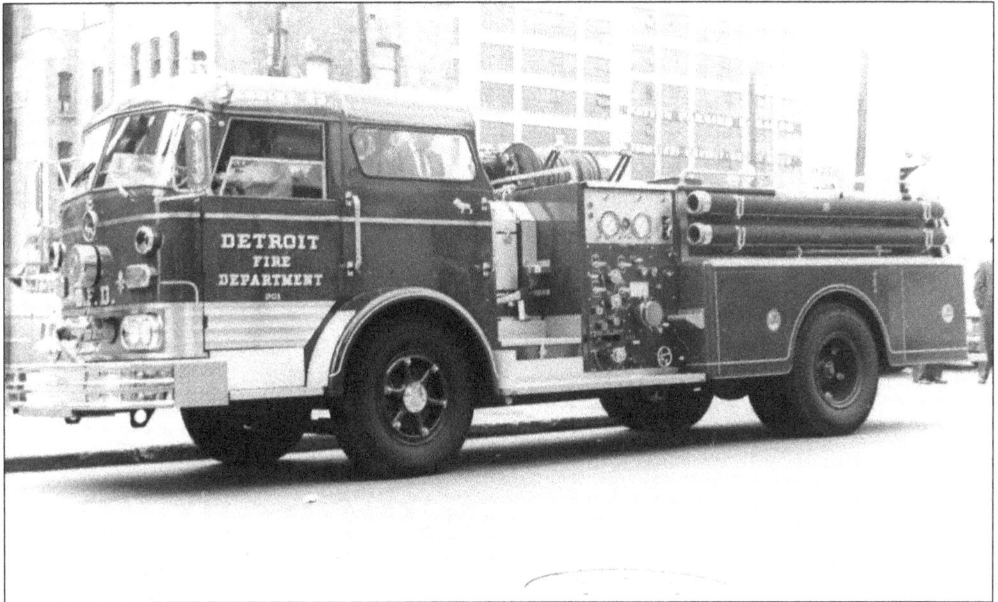

Engine Company No. 23's Mack cab over an engine pumper was located at East Grand Boulevard and Moran Street on Detroit's east side. (Courtesy of Matt Lee.)

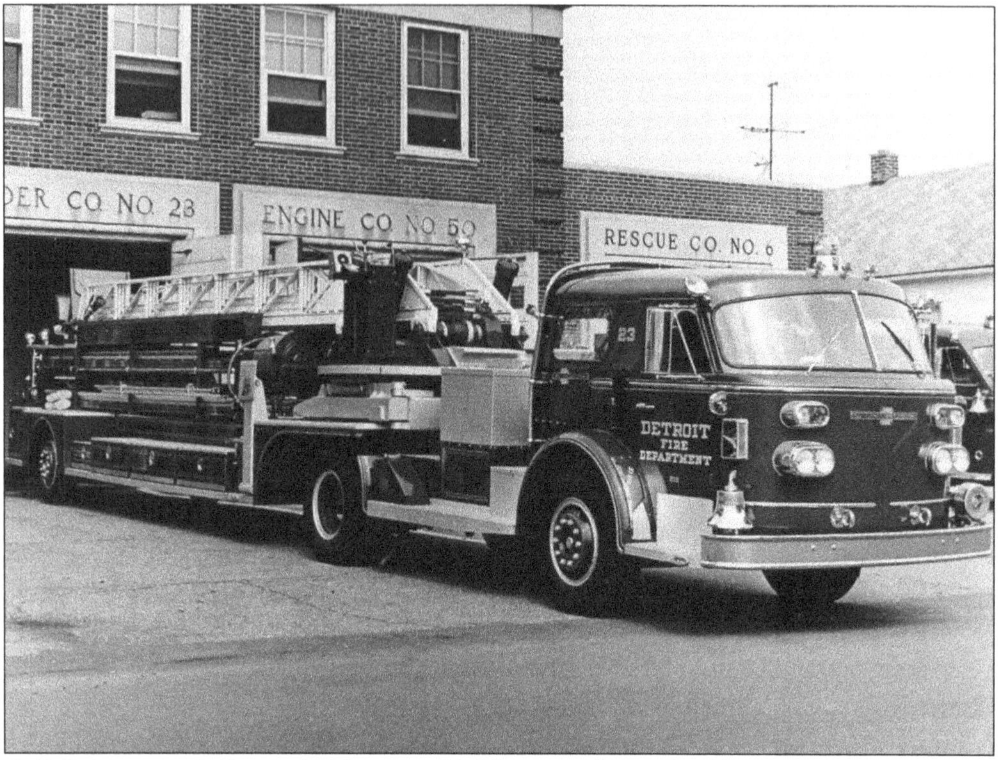

A 1969 American LaFrance 100-foot aerial ladder sits in front of Ladder Company No. 23. This two-piece vehicle was referred to as a "tiller truck." The features of this truck include a booster tank with high-pressure fog reels. Ladder Company No. 23 shared the firehouse with Engine Company No. 50, located at Houston Whittier Street. Later, this firehouse became the center of attention in the fire documentary *BURN*. (Courtesy of Matt Lee.)

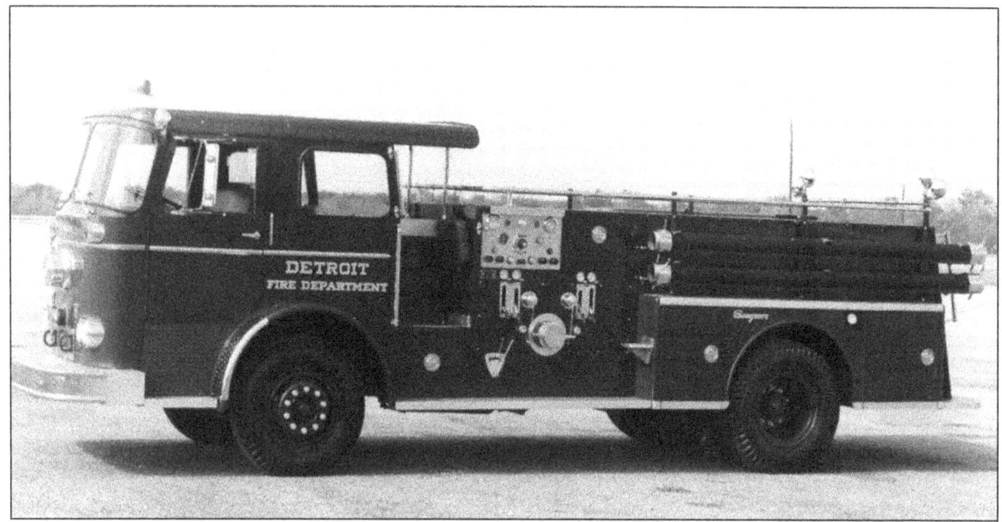

Purchased in November 1964, this unusual Seagrave canopy cab over pumper, which was assigned to Engine Company No. 5, was the only one of its type in the DFD. The canopy-roof style was designed to protect firemen riding on the engine from getting hit with objects thrown at them and shield them from the weather. (Courtesy of Matt Lee.)

Members of the Detroit Fire Department pose for a group photograph at a multiple-alarm fire by the Salvation Army Canteen Unit. Note the rubber coats and plastic helmets with metal shields. The Salvation Army has an exclusive Canteen Unit assigned to support the DFD for any major incidents. (Courtesy of Matt Lee.)

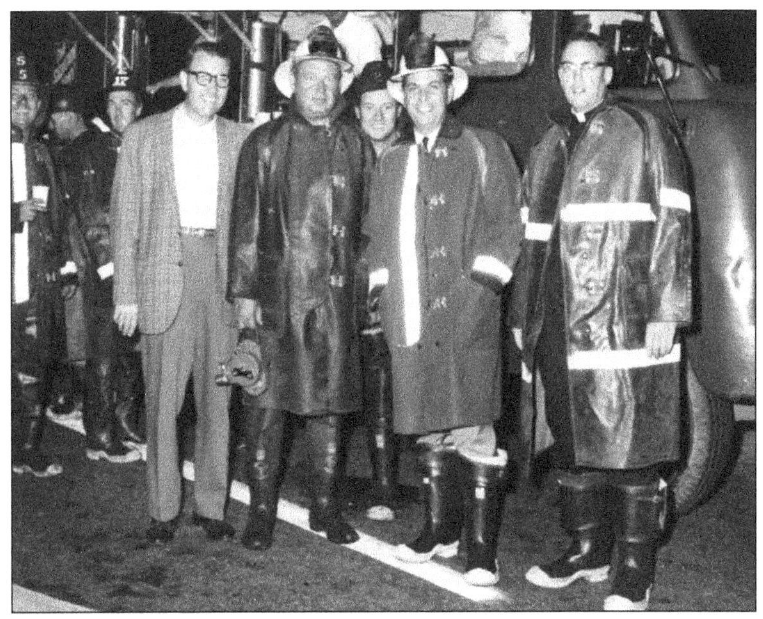

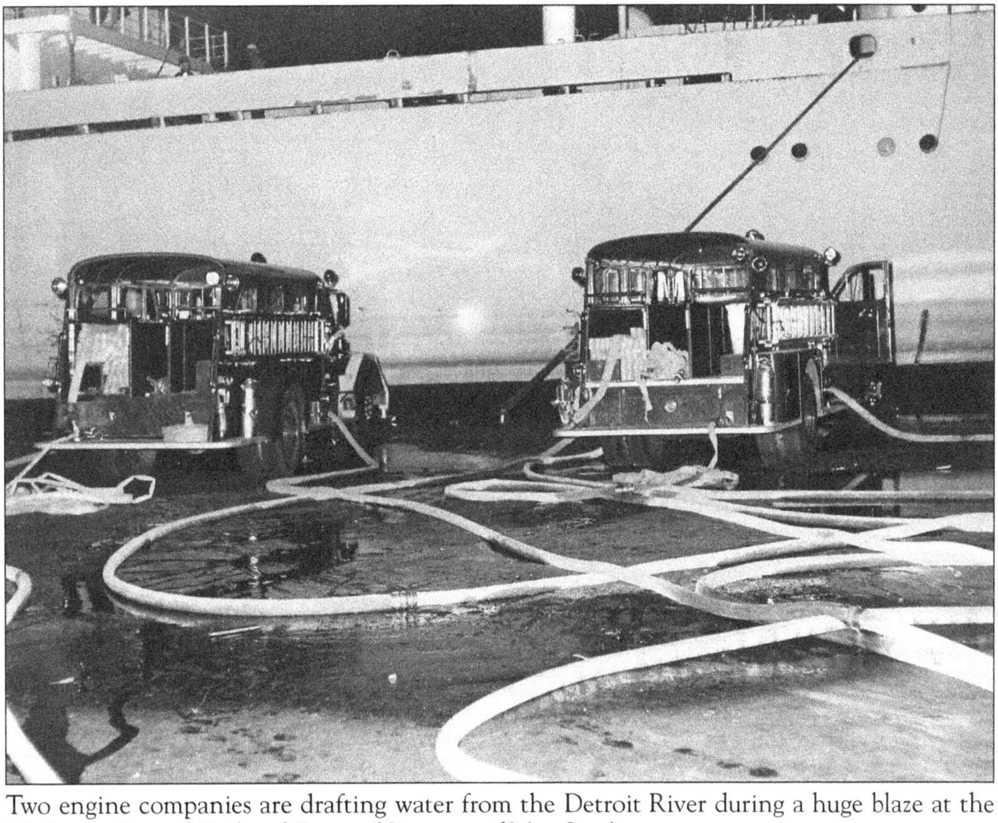

Two engine companies are drafting water from the Detroit River during a huge blaze at the Michigan Central Railroad Depot. (Courtesy of Matt Lee.)

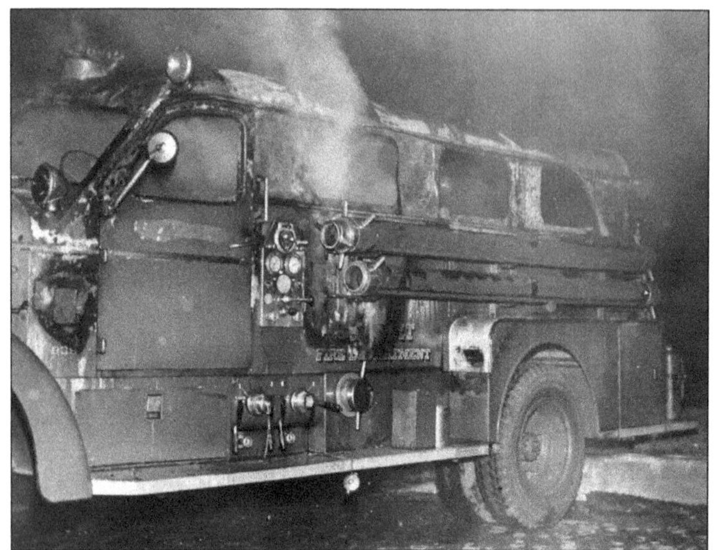

This close-up view of Engine Company No. 8 shows the damage to a 1952 Seagrave sedan after catching fire from radiant heat while connected to a hydrant directly in front of the Michigan Central Railroad Depot. Due to a sudden back draft and searing flames, the crew was unable to disconnect and move the rig to safety. The five-alarm fire took place on June 19, 1966. (Courtesy of Matt Lee.)

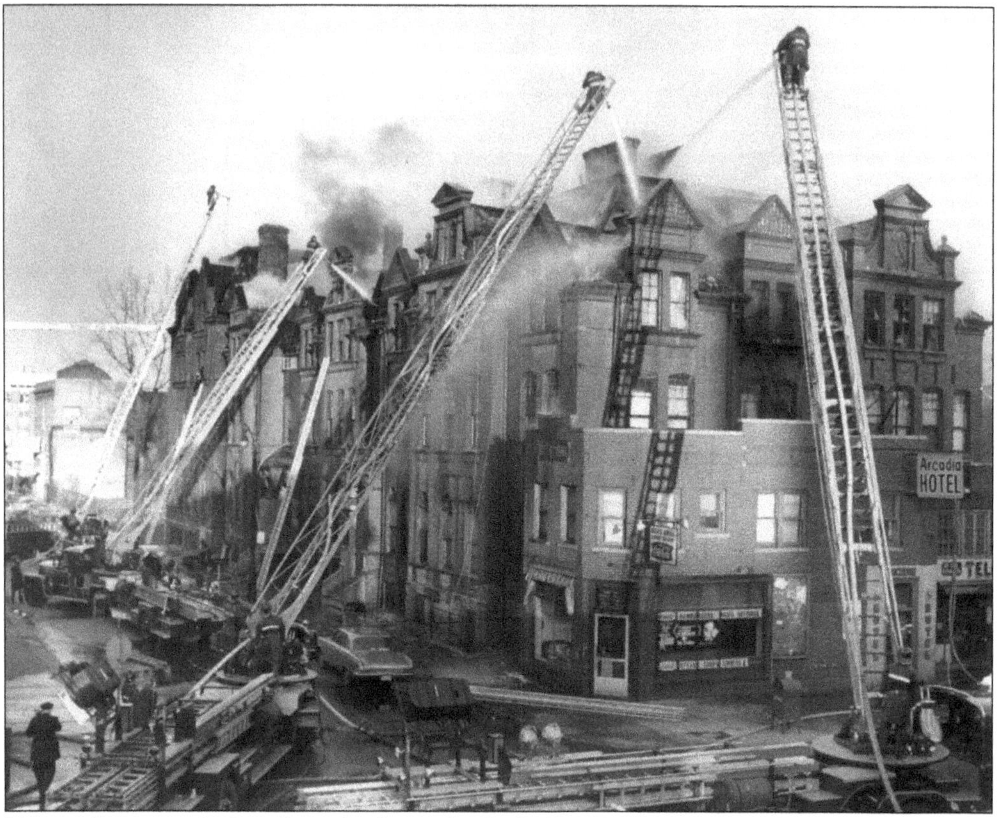

Here is the scene of a five-alarm fire located at 3468 Cass Avenue corner of Stimson Street. The blaze took place on March 1, 1966, at box No. 669. Extra companies were requested beyond the five alarm. The Arcadia Hotel, the former Liggett School building, was the site of the fire. One guest died in the center top floor. Multiple aerial ladders can be seen operating master streams, and a snorkel is operating far off in the distance as well as portable ground ladders raised to the upper floors. (Courtesy of Matt Lee.)

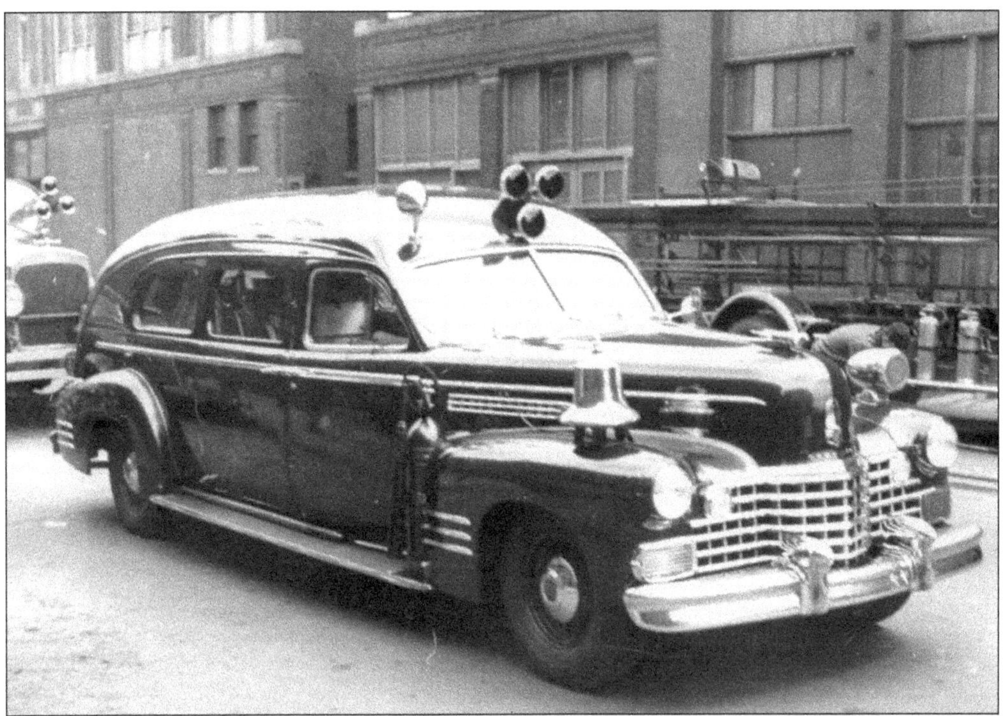

Here is a Cadillac-made Detroit Fire Department ambulance, and faintly visible on passenger side door are the words "Detroit Fire Dept. Rescue 3." The Detroit Fire Department had two ambulances; Rescue No. 4 was the other one. (Courtesy of the Detroit Historical Society.)

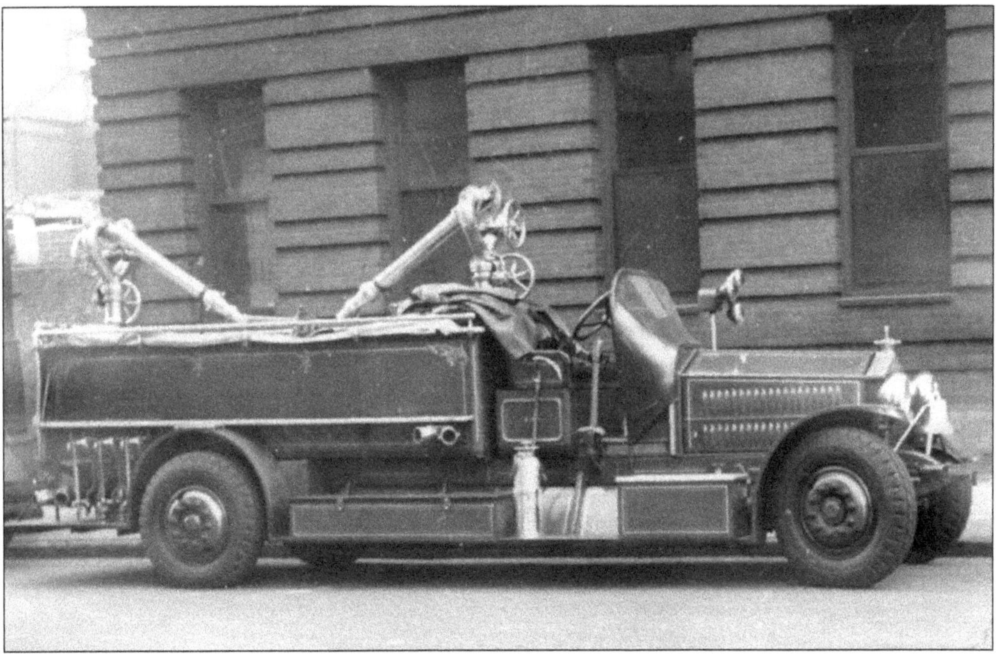

The Detroit Fire Department high-pressure wagon, a 1922 Ahrens-Fox, is seen parked around 1950. Two high-pressure nozzles, in lowered position, are visible. The vehicle is right-hand drive with the steering wheel on the right side. (Courtesy of the Detroit Historical Society.)

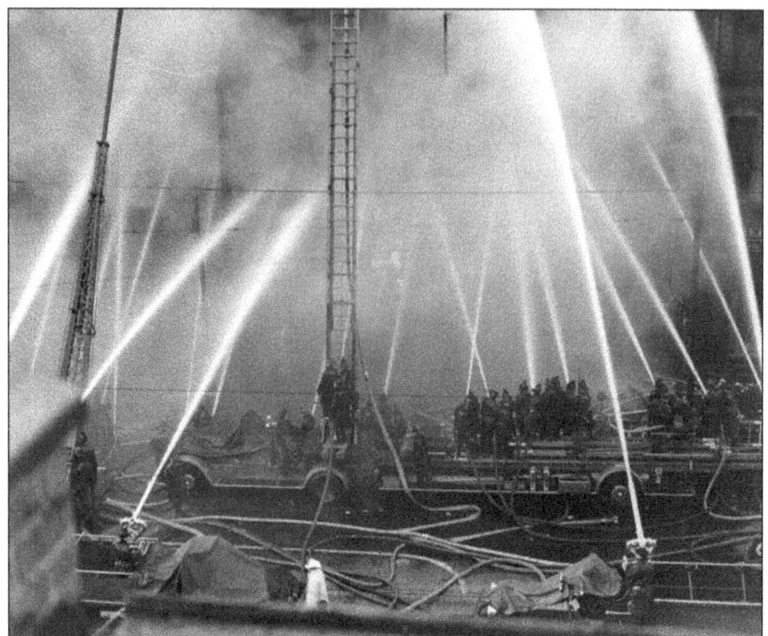

This raging multiple-alarm fire has firemen in a defensive stand. In Detroit, men and women of the DFD come to work to do their job on a daily basis, and many of the fire scenes have become routine, meaning the different fires start to blend together. (Courtesy of David Traiforos.)

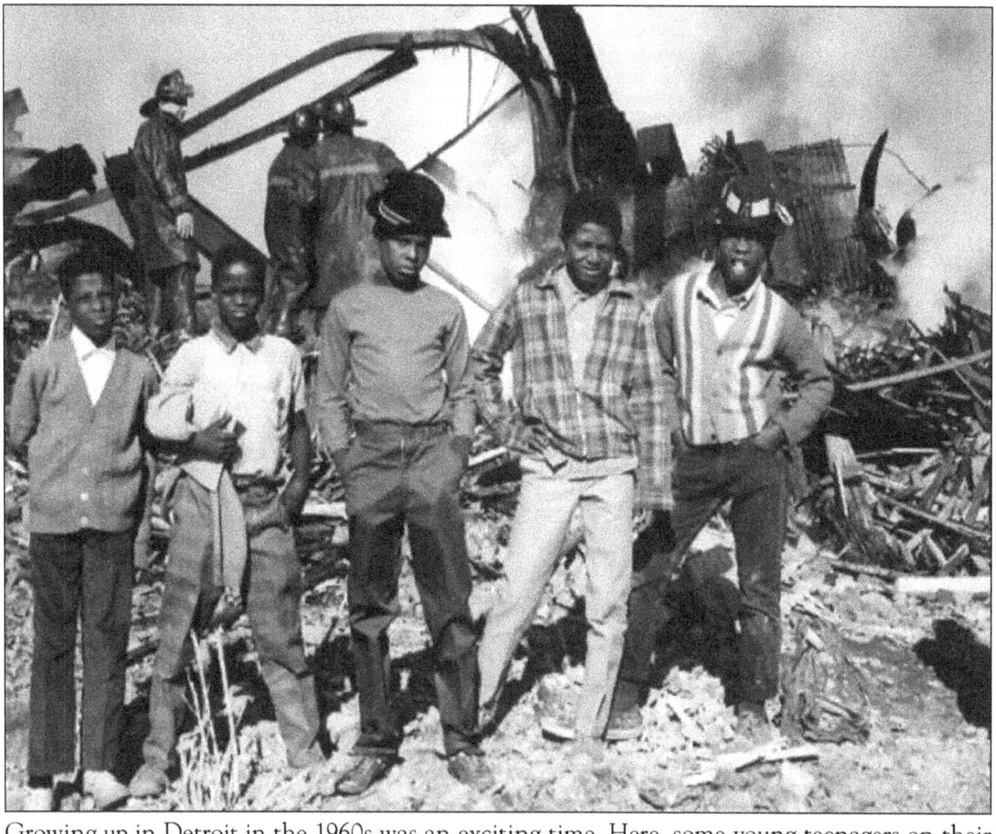

Growing up in Detroit in the 1960s was an exciting time. Here, some young teenagers on their way home from school stop to investigate the scene of another house fire in their neighborhood. (Courtesy of Bill Eisner.)

Members of Engine Company No. 27 operate a hose line in a defensive posture at a multiple-alarm fire. As the steady decline in population in Detroit took place, fire companies started to see an increase in fire activity. (Courtesy of Bill Eisner.)

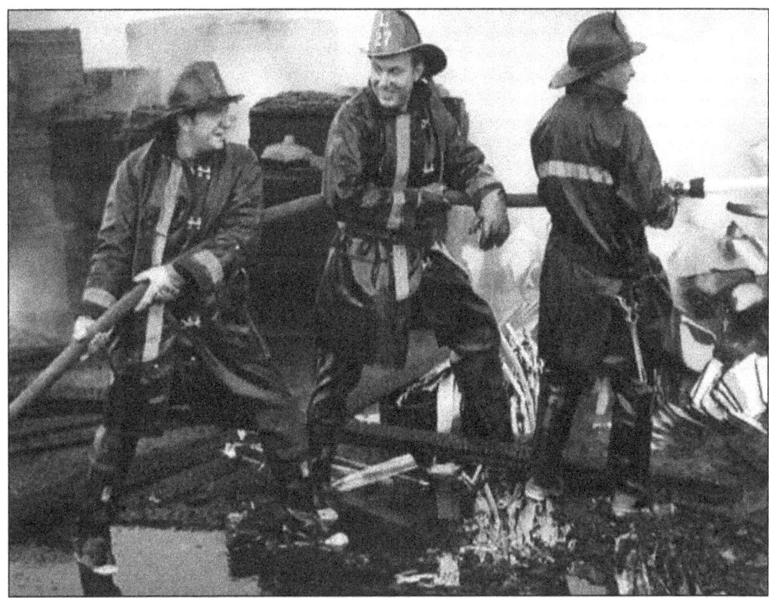

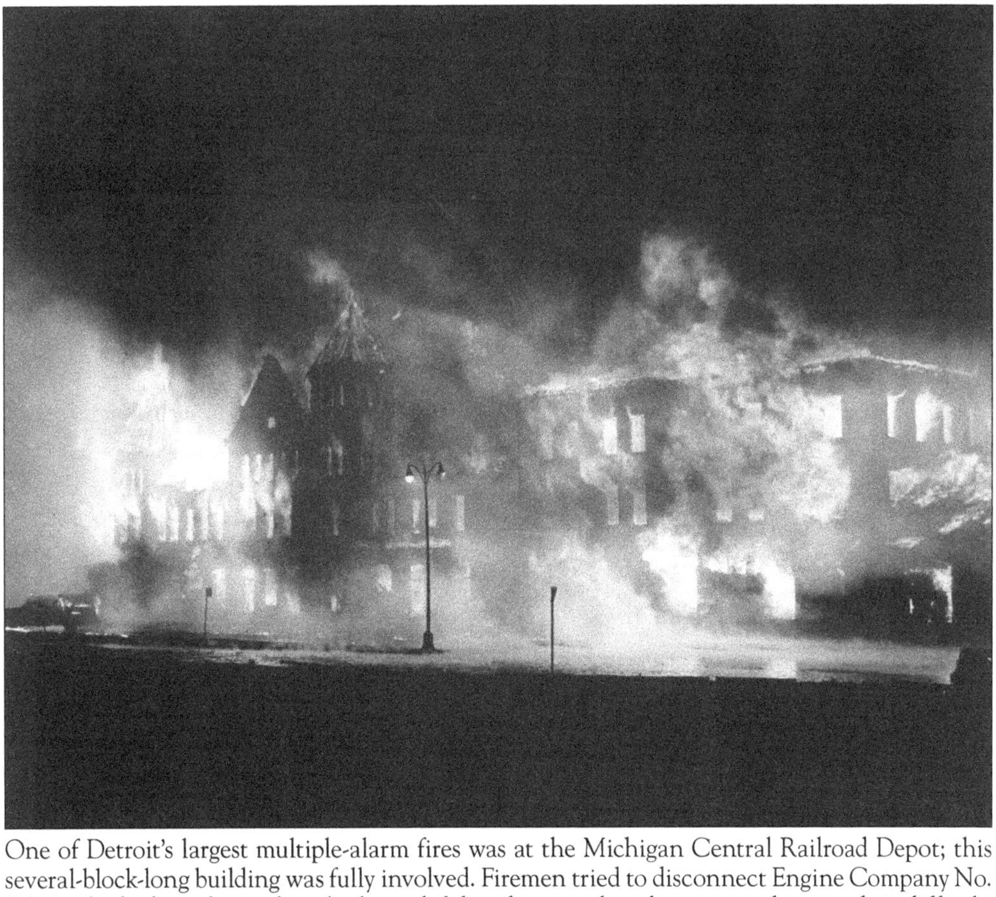

One of Detroit's largest multiple-alarm fires was at the Michigan Central Railroad Depot; this several-block-long building was fully involved. Firemen tried to disconnect Engine Company No. 8 from the hydrant located in the lower left-hand corner, but the extreme heat made it difficult, so the engine caught fire and burned. (Courtesy of Jon Soave.)

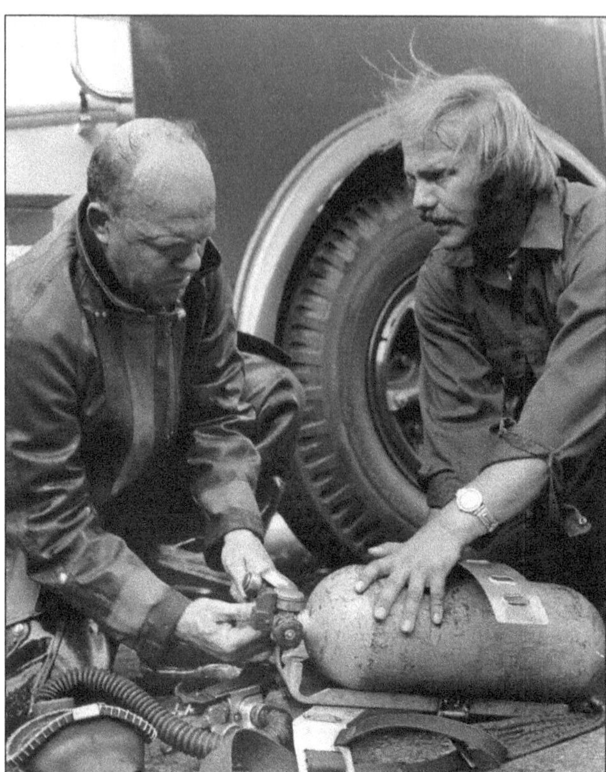

Here, a driver assists a fireman with changing his self-contained breathing apparatus bottle so he can prepare to re-enter the building for another round of hard work. (Courtesy of Jon Soave.)

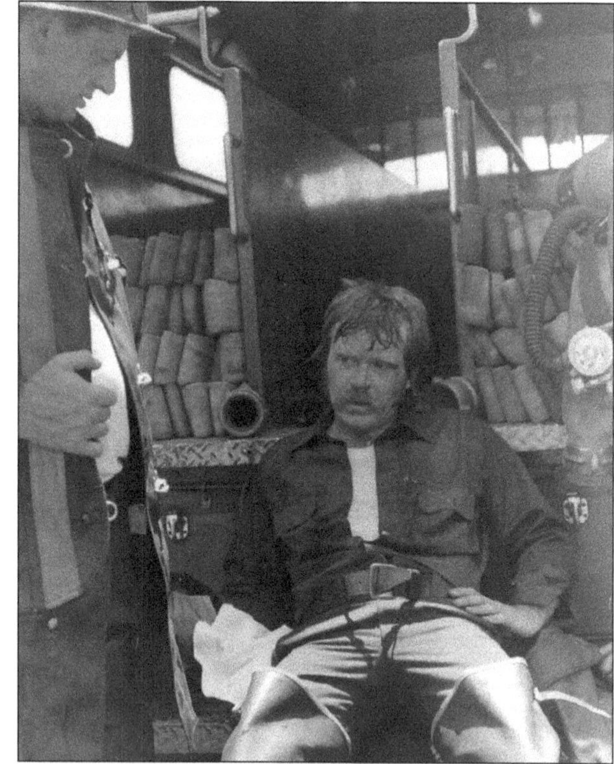

The sheer look of exhaustion is shown on this Detroit fireman's face while he attempts to catch his breath and takes a break sitting on the back step of this Seagrave sedan pumper. (Courtesy of Jon Soave.)

A Detroit fireman is dragged out of a burning building and attended to by other firemen. (Courtesy of Jon Soave.)

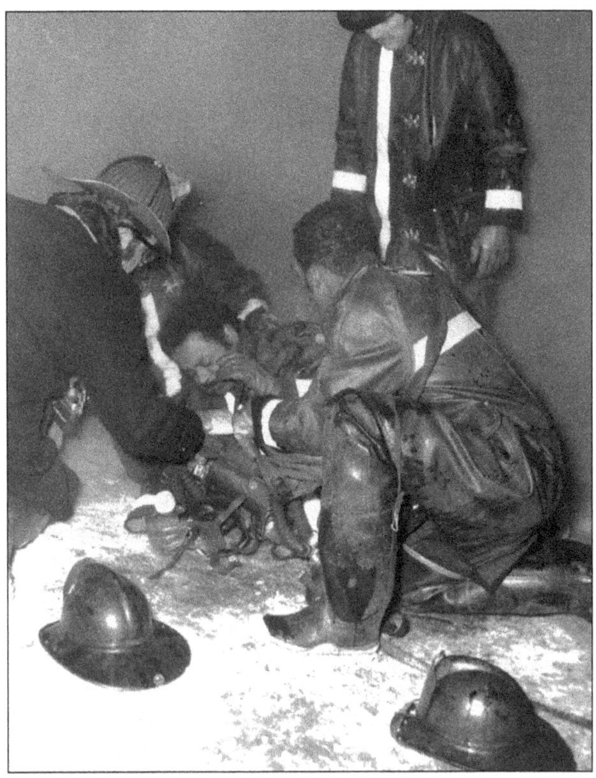

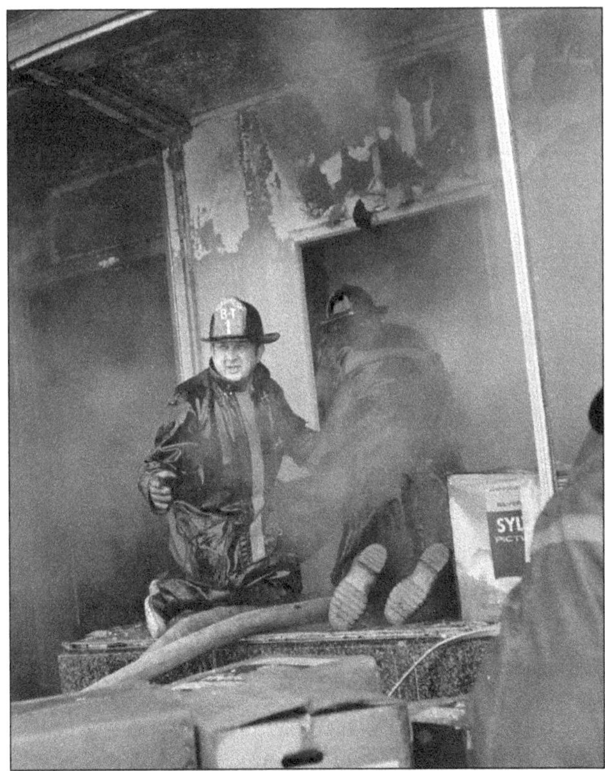

Here, a sergeant, assigned to Fireboat Tender No. 1, shouts out for more hose so other firemen and the nozzleman can advance the hose farther into this commercial building to reach the seat of the fire. (Courtesy of Jon Soave.)

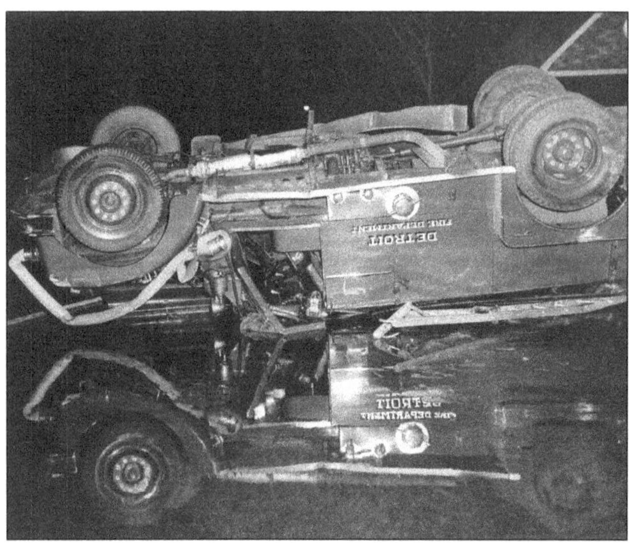

On November 9, 1963, this rollover accident involved Engine Company No. 56, which was responding to a street corner box at Hildale Avenue and Conely Street. Fireman Albert J. Booth Jr. was killed, and another fireman was seriously injured. The significant damage shows the roof of the enclosed vehicle completely crushed. The Seagrave sedan had the enclosed style, which may have made this accident worse. A 15-year-old boy was caught by police and admitted to turning in the false alarm. (Courtesy of Jon Soave.)

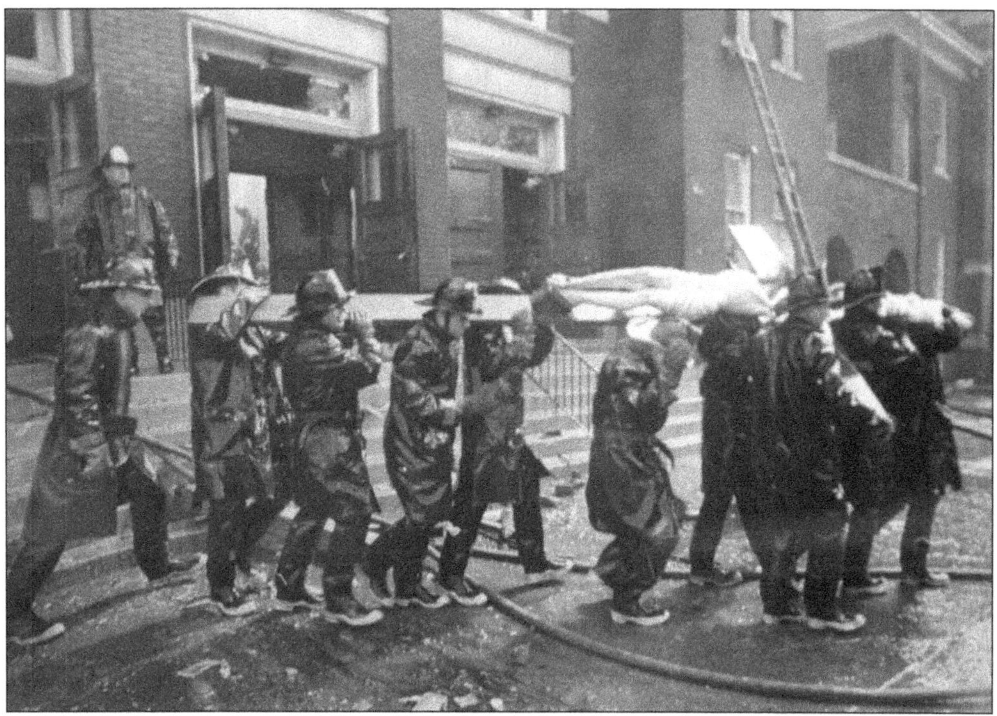

On April 10, 1963, a multiple-alarm fire destroyed the Our Lady of Sorrows Catholic Church, located at Meldrum Street and Mack Avenue on Detroit's east side. This moment was captured when Detroit firemen carried out the church's crucifix. The cross was stored for many years, but in 2011 it was hung again at St. Margaret of Scotland Church in St. Clair Shores, Michigan. (Courtesy of Bill Eisner.)

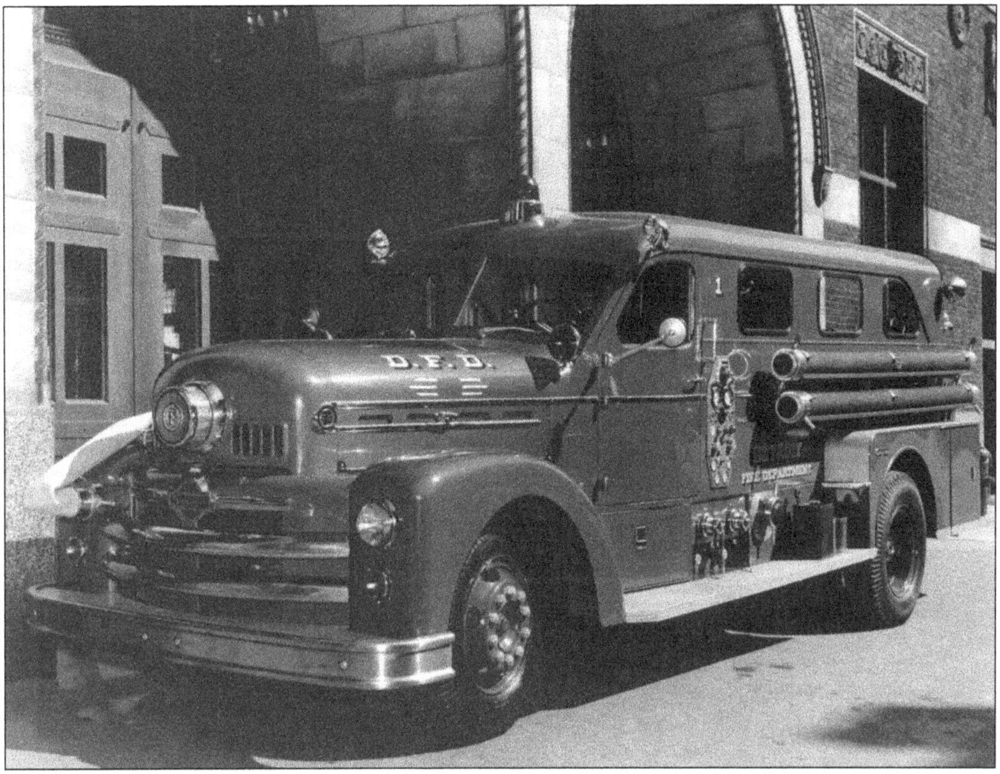

Engine Company No. 1 is parked alongside headquarters on the Washington Boulevard side of the street. This is a classic firehouse with arched, overhead swing-out wooden doors. (Courtesy of Jon Soave.)

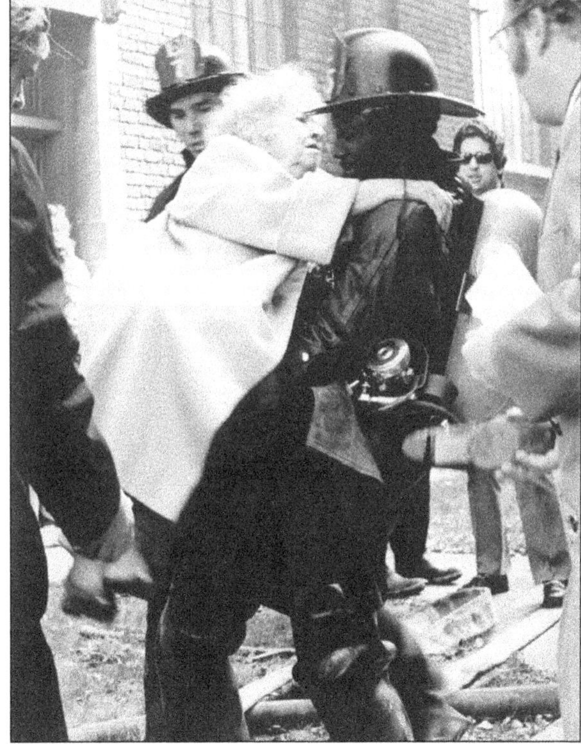

Firemen are trained to search for victims in fires and rescue them if they find anyone trapped or overcome by smoke. Here, a fireman carries a woman he just rescued from a fire in her apartment, while other members come to his aid. (Courtesy of Jon Soave.)

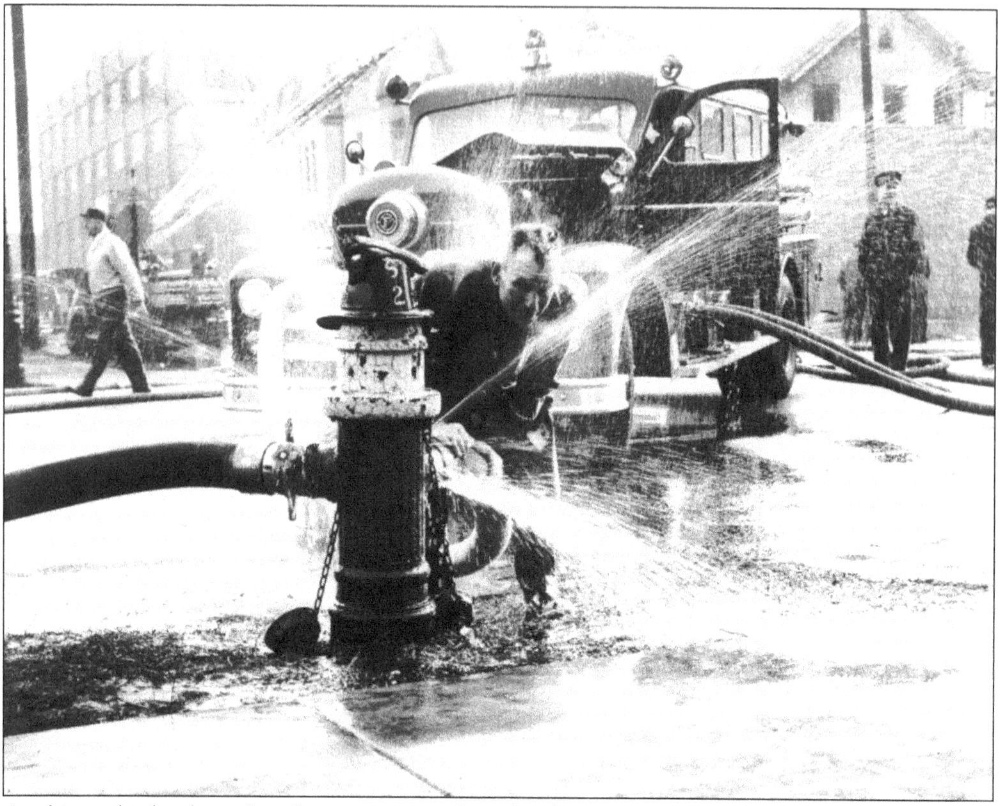

At this multiple-alarm fire, firemen are working hard; one is pictured cooling off and grabbing a drink of water at a hydrant after having exerting himself helping to control the fire. (Courtesy of the Joe Mancinelli family.)

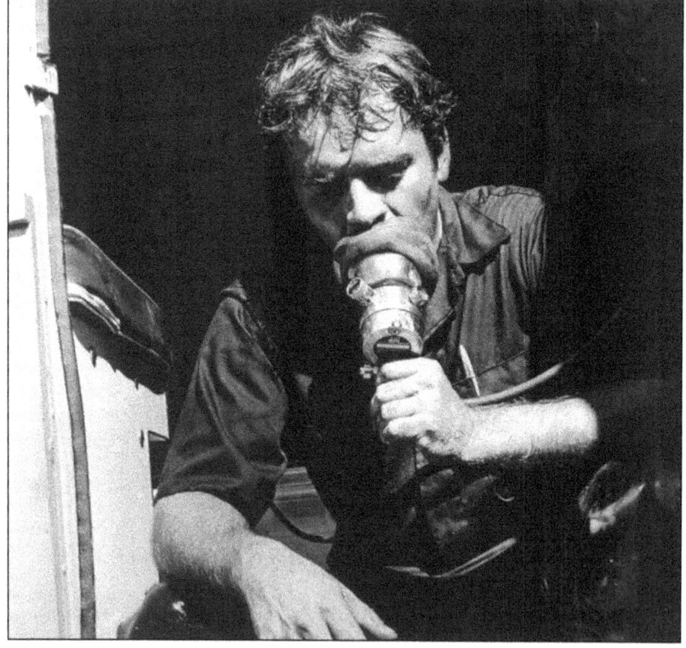

Here, a fireman has to self-administer oxygen after working hard at a fire; his basic duties are to extinguish, overhaul, and ventilate. On hot, humid summer days, firemen can tire out due to the extreme heat put on the body, the number of fires they have responded to that day, and the lack of manpower if a fire is understaffed. (Courtesy of Jon Soave.)

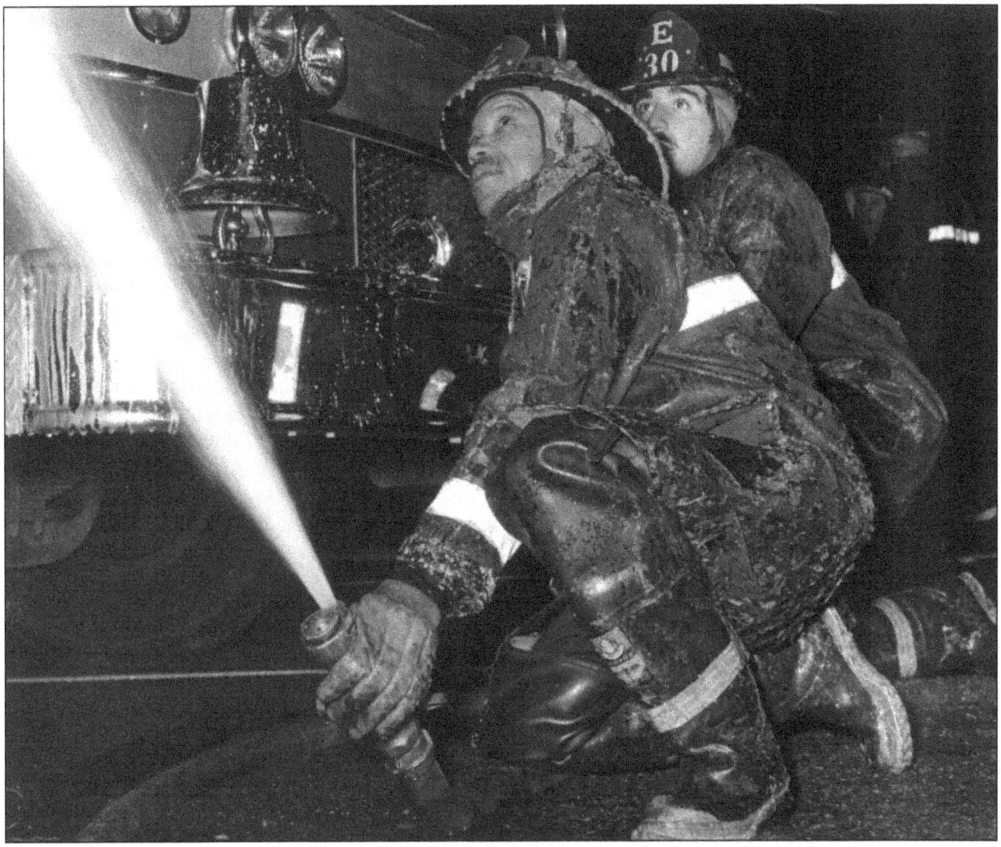

On a cold night in December in the 1960s, members of Engine Company No. 30 operate a hose line with a Rockwood nozzle on it. The Rockwood nozzle was the preferred choice of the DFD for many years. DFD firemen wore plastic MSA (mine safety appliance) helmets with Detroit shields, rubber coats with reflective trim, Servus three-quarter pull-up fire boots, and Red Ball fire gloves. (Courtesy of Jon Soave.)

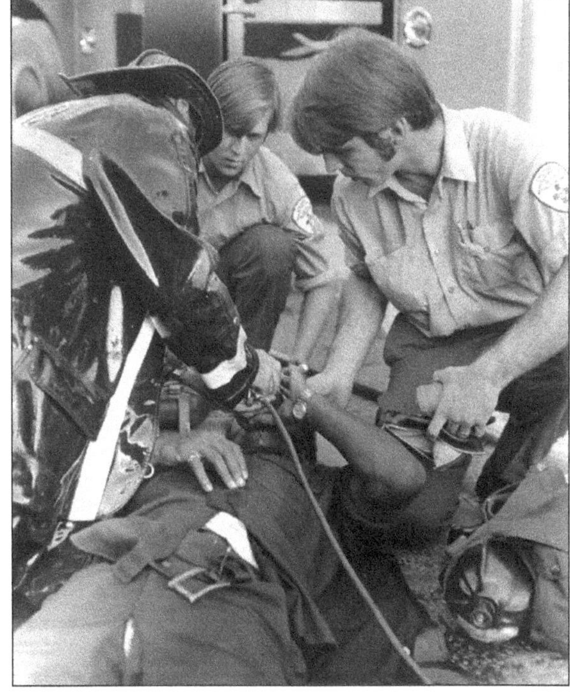

Working at fires continuously without a break can cause firemen to become fatigued and exhausted. Here, members of Detroit EMS and firemen administer oxygen to a fireman suffering from smoke inhalation and exhaustion while working at a fire on a warm day. The turnout gear helps add issues on warm days. (Courtesy of Jon Soave.)

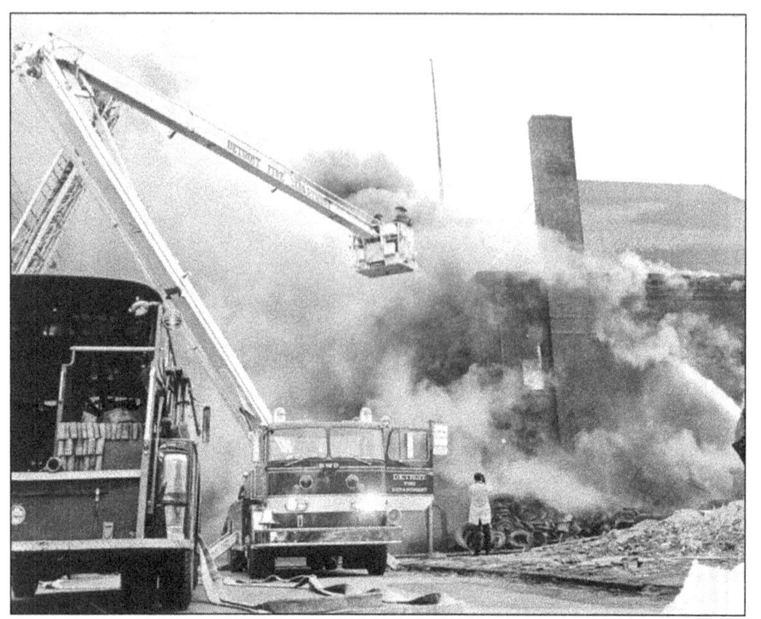

Detroit Snorkel Company No. 1 prepares to go to work at this multiple-alarm fire in a commercial building. The FWD snorkel operates a two-section boom device with a reach of 90 feet. A basket is attached at the end of the boom to hold two to three firemen and to flow water from a prepiped water system. (Courtesy of Jon Soave.)

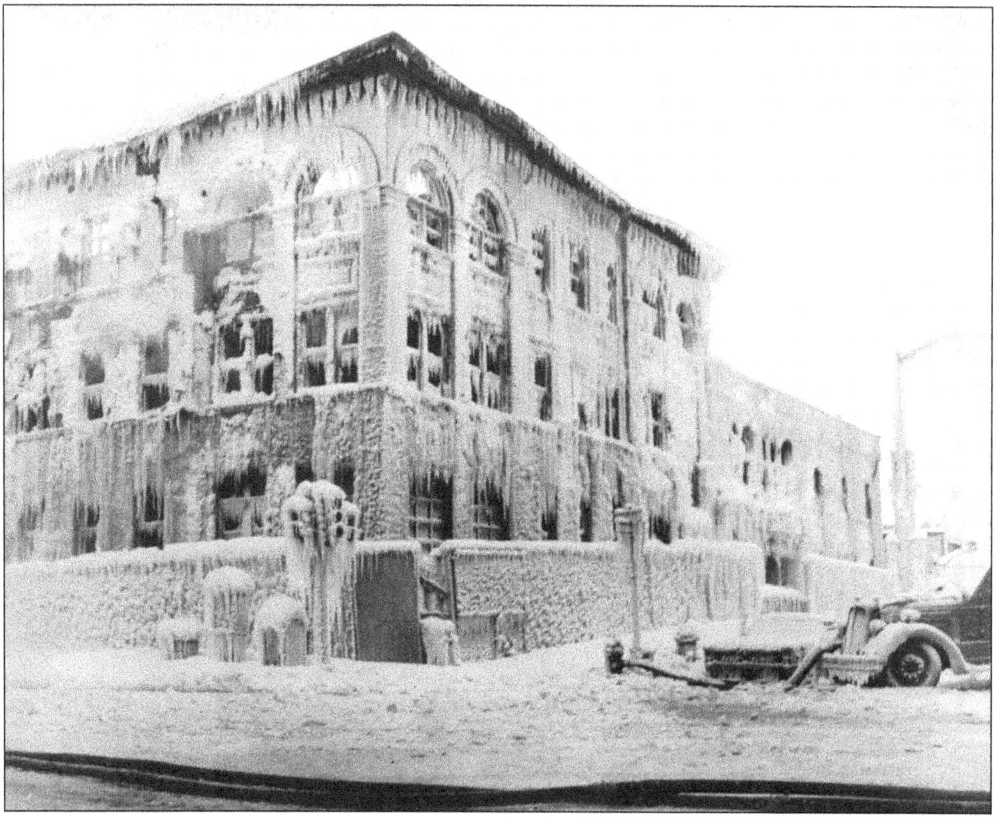

Cold-weather fires with temperatures below zero usually create images of ice castles on the burning building as thousands of gallons of water freeze. Note the amount of ice on the traffic light on the corner. Even the fire engine is frozen to the hydrant. (Courtesy of Jon Soave.)

Five

THE 1967 DETROIT RIOTS

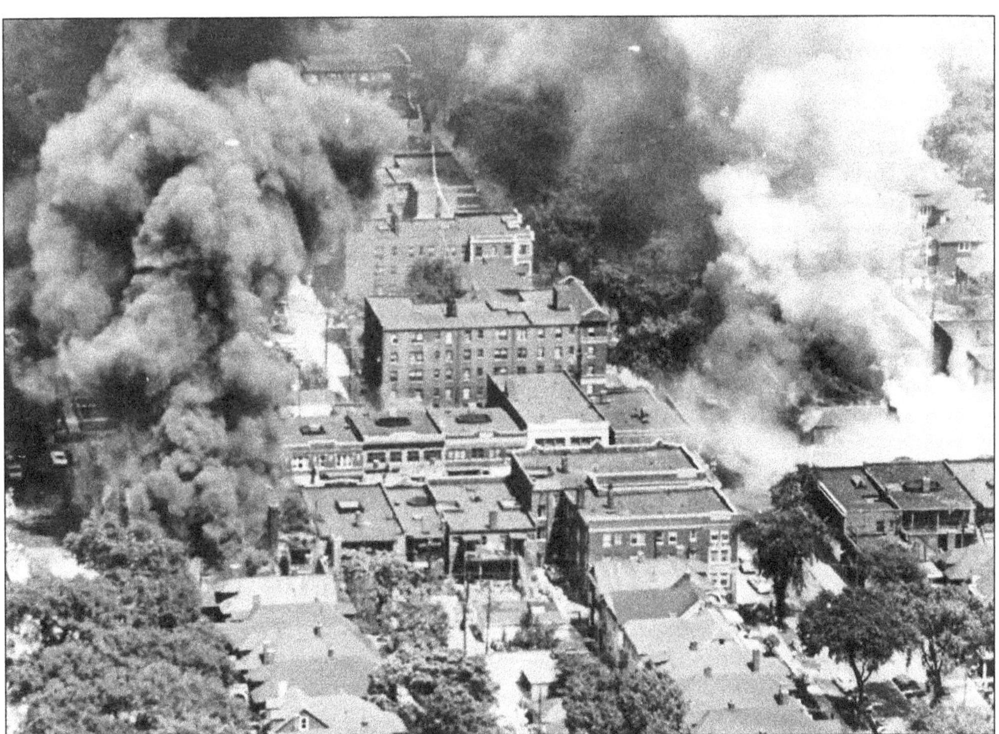

Here is an aerial view of the city of Detroit; it shows different fires on Sunday, July 23, 1967. For the first time in the city's history, the fire department's telegraph system tapped out the ominous signal 3-777, which is "Recall All Firefighters, Detroit is Burning." (Courtesy of the Detroit Firemen's Fund.)

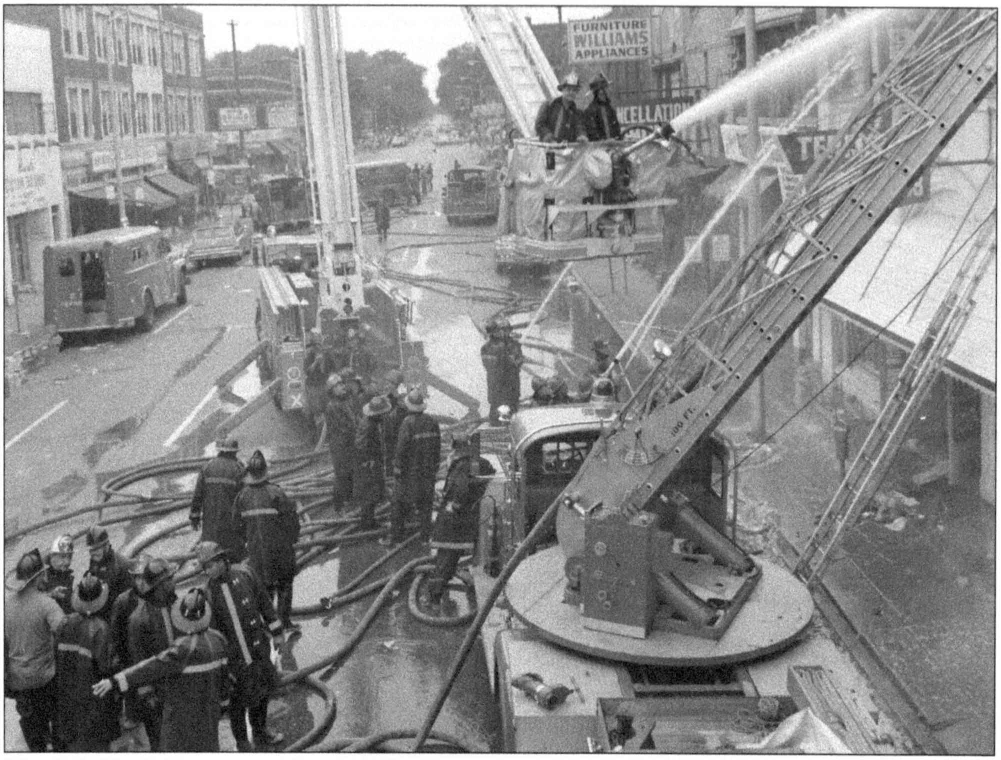

This July 23, 1967, scene is of an early Sunday morning fire at Twelfth Street between Taylor Road and Clairmont Street. The blaze was started in the Cancellation Shoe store. It was the result of looting and arson. Here, firemen had to work together in close proximity to one another to avoid being outside of the safety zone so as not to be a target to be shot or have objects thrown at them. Police personnel set up safe areas while watching over them with guns drawn. (Courtesy of Matt Lee.)

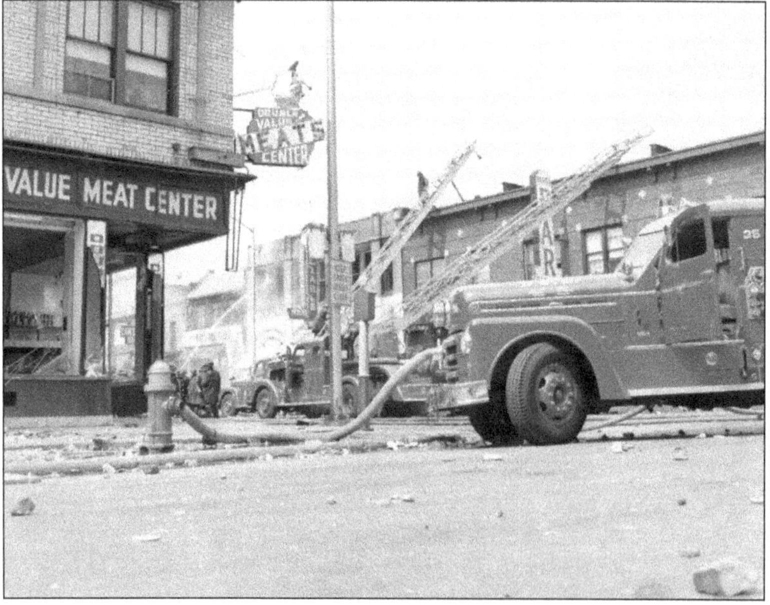

The "Summer of 67," also referred to as the Detroit riots/civil unrest, started July 23 and lasted until July 27. Here, DFD units operate at a fire in a commercial building. The debris in the streets shows what objects were picked up and thrown at DFD firemen. (Courtesy of Matt Lee.)

Having items thrown at them, firemen call the police for help. Police resources were stretched thin with the number of incidents taking place in the city, but when the situation was reorganized, police units responded to fire incidents and were eventually assigned to fire task force units. (Courtesy of the Detroit Firemen's Fund.)

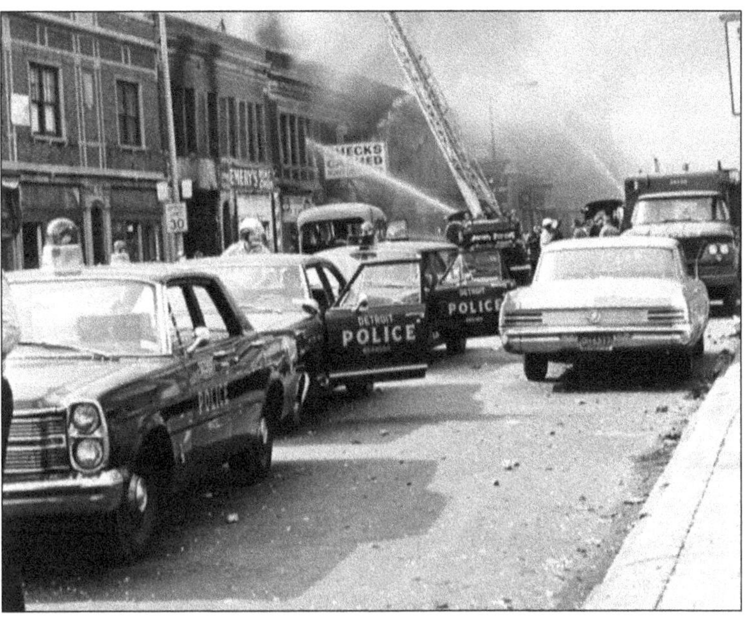

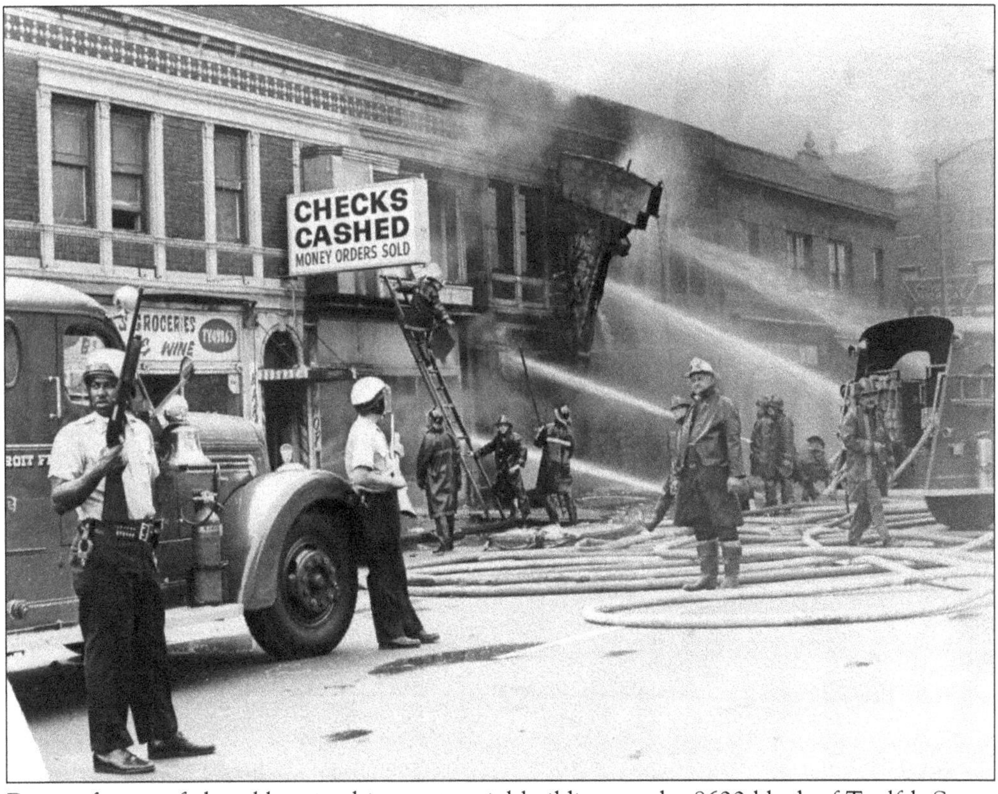

Detroit firemen fight a blaze in this commercial building on the 8600 block of Twelfth Street under the protection of armed Detroit Police officers. They came under attack when looters threw objects and fired guns at them. (Courtesy of the Detroit Firemen's Fund.)

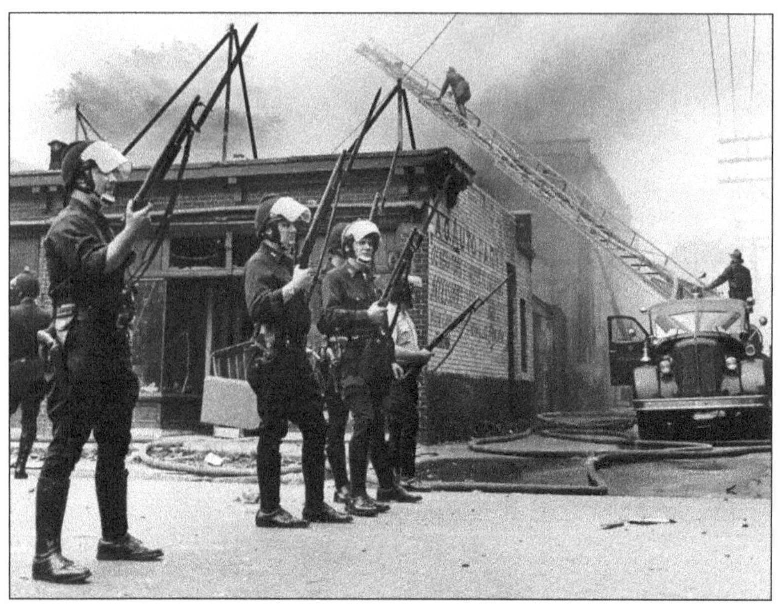

Due to the violence becoming more serious in nature, police had to don riot gear. Now, when police stood by to protect members of the DFD, their own protection was stepped up a notch to defend all public safety servants working. (Courtesy of the Detroit Firemen's Fund.)

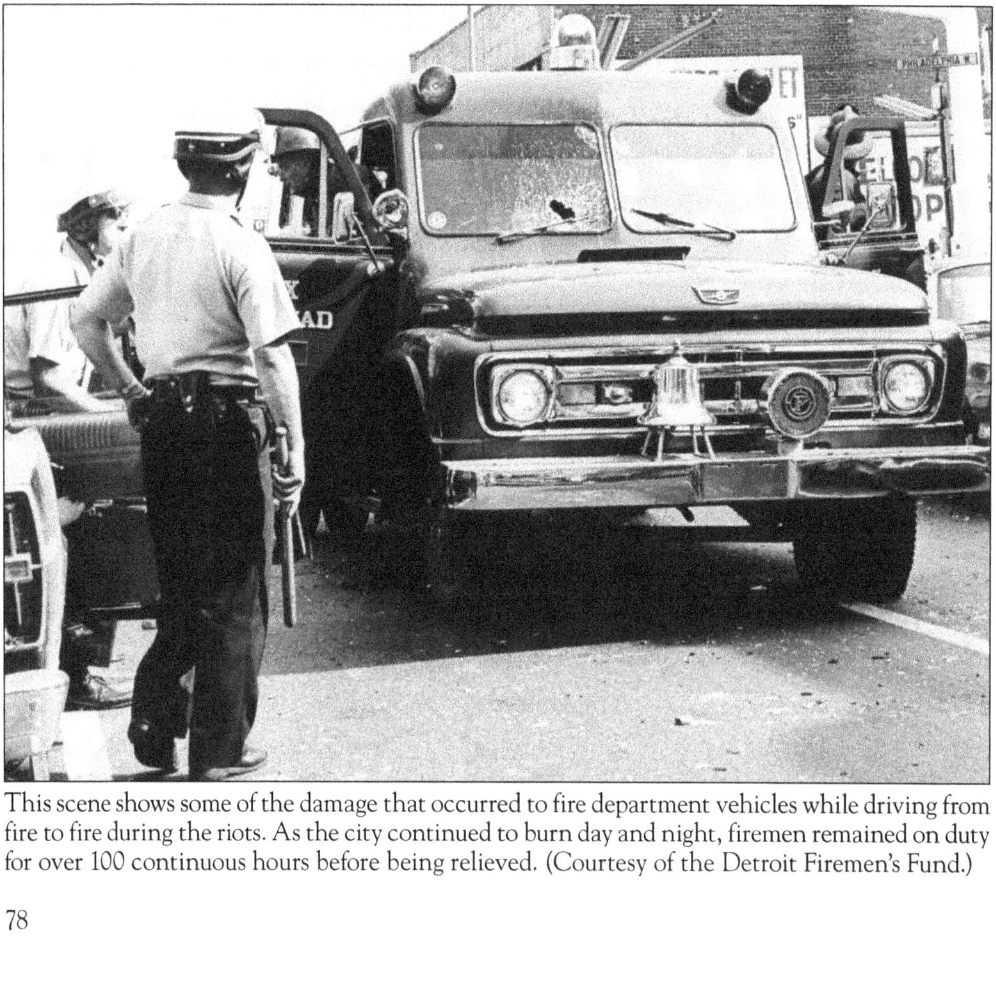

This scene shows some of the damage that occurred to fire department vehicles while driving from fire to fire during the riots. As the city continued to burn day and night, firemen remained on duty for over 100 continuous hours before being relieved. (Courtesy of the Detroit Firemen's Fund.)

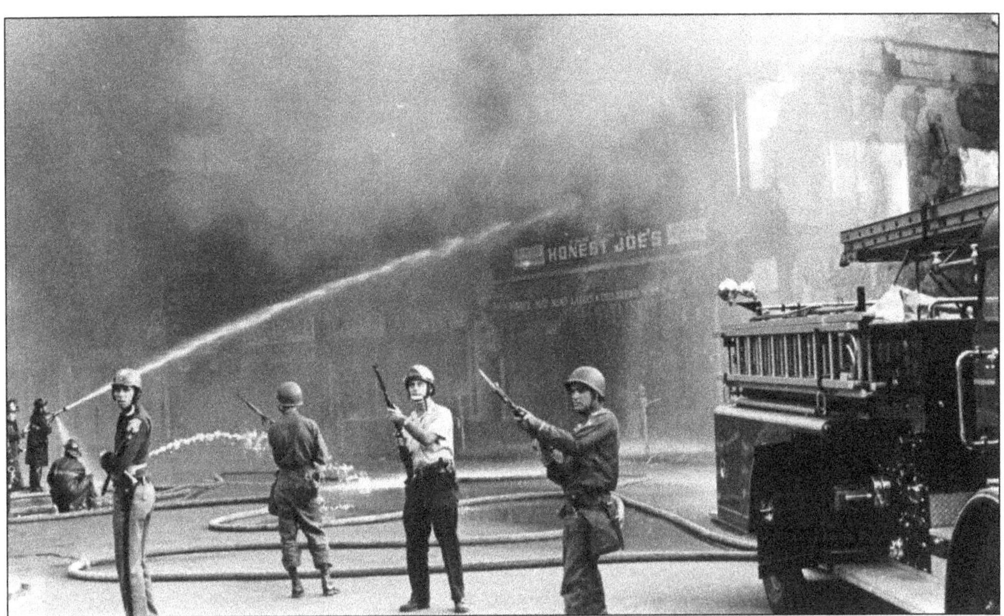

The City of Detroit realized that the number of incidents was growing hour by hour in the first couple of days of the riots, so a call was made to Gov. George Romney of Michigan for additional resources. The Michigan National Guard sent in local troops, and Pres. Lyndon Johnson sent in federal troops from the 82nd Air Borne Division. Here, Detroit Police and National Guard troops stand by with drawn weapons to protect firemen performing their duties to save lives and extinguish fires. (Courtesy of Detroit Firemen's Fund.)

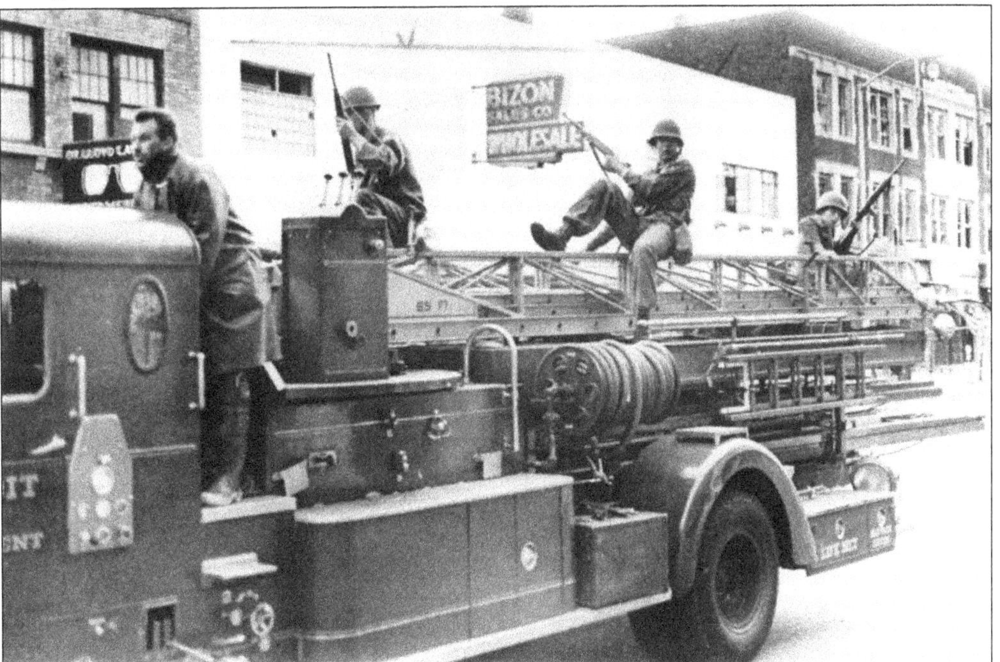

Here, National Guardsmen ride on the aerial ladder of a ladder truck responding to a fire. The apparatus could only seat five members, so the soldiers had to ride the rig, where they would be easy targets. (Courtesy of the Detroit Firemen's Fund.)

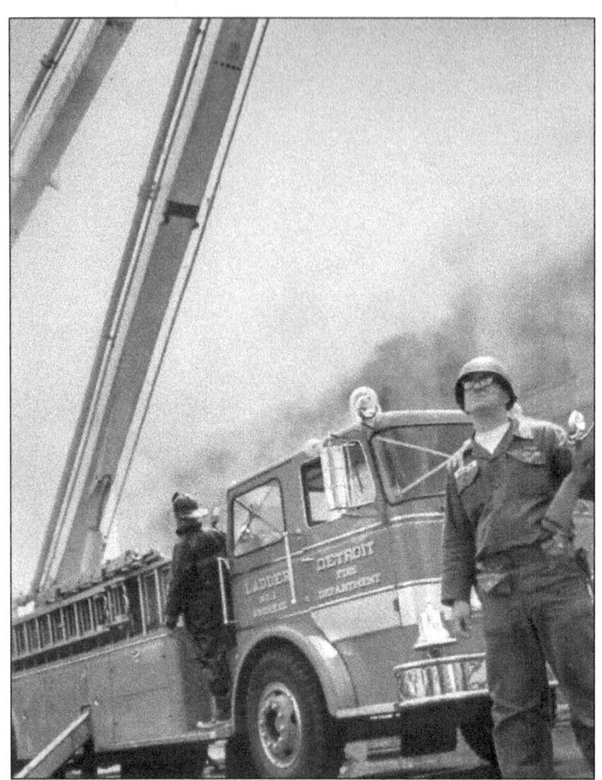

A National Guard soldier stands as a lookout for people either throwing objects or shooting guns at firemen. The riots lasted until July 27. Two Detroit firemen, Carl E. Smith and John Charles Ashby, died of injuries sustained during the civil unrest; one Detroit police officer was killed. (Courtesy of the Joe Mancinelli family.)

Here, a ladder truck responds to another fire during the civil unrest period. The tiller man driving the rear of the ladder truck is surrounded by soldiers. (Courtesy of the Joe Mancinelli family.)

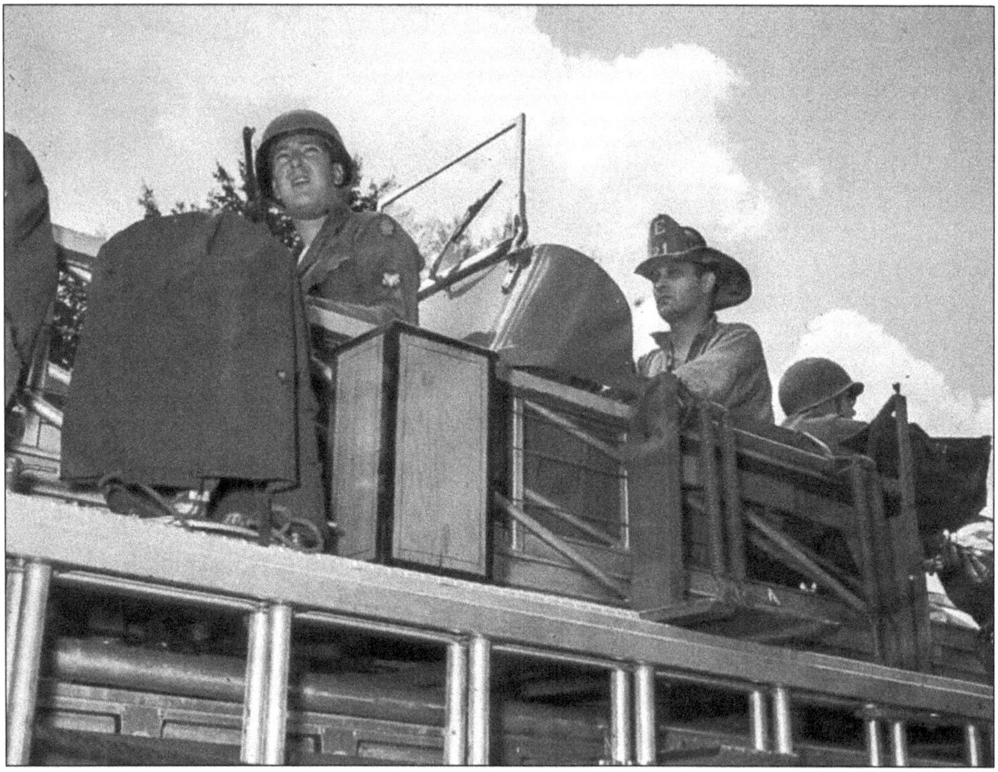

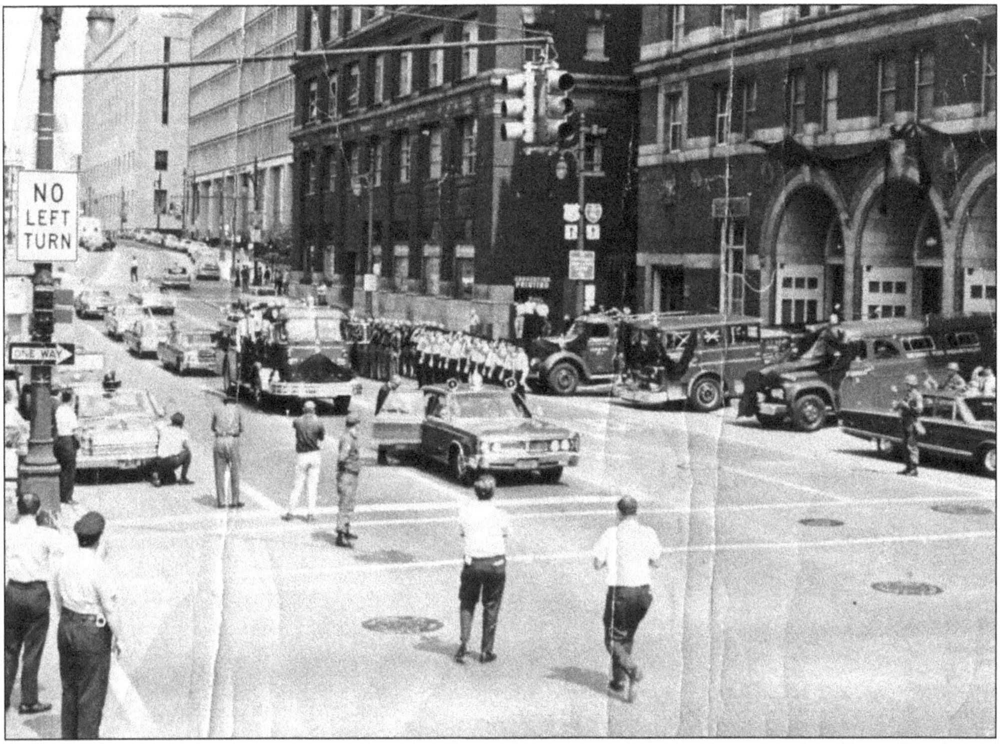

This photograph was taken at the Detroit Fire Department Headquarters; the flag-draped casket of Fireman Carl E. Smith, who was assigned to Ladder Company No. 11, died of a gunshot wound at a fire scene. His funeral took place during the riots as soldiers and police stand by to help prevent further disturbances. Fire apparatuses parked on the aprons are draped in bunting. Firemen and soldiers standing at attention give a final salute to Fireman Smith as his casket on the fire engine drives by. (Courtesy of the Detroit Firemen's Fund.)

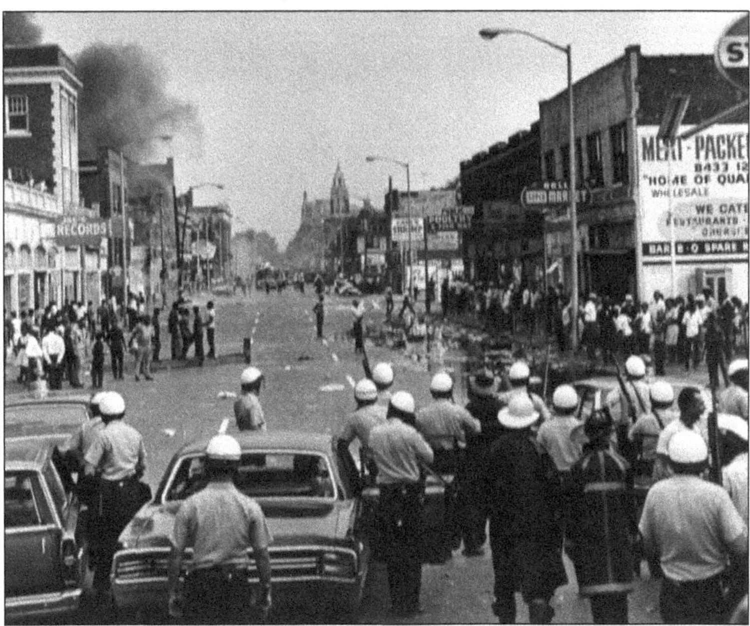

The confrontations continued to take place for several days as did the fires, which were caused by arson. There were 1,682 fire runs, where 1,300 buildings were destroyed; 2,700 businesses were looted; 5,000 Detroit residents were left homeless after the riots. (Courtesy of the Joe Mancinelli family.)

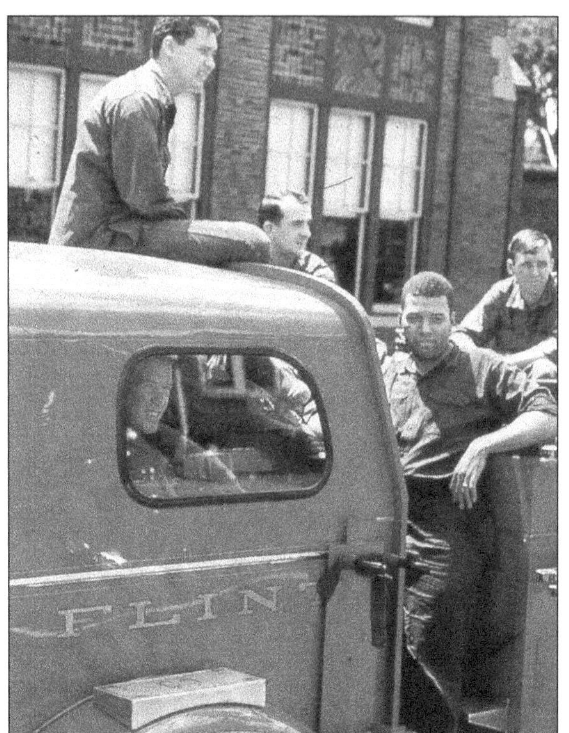

Members of the Flint Fire Department stand by for an assignment in the staging area for fire equipment. Flint is located 75 miles north of Detroit and was one of 44 fire departments that responded to aid Detroit. (Courtesy of the Joe Mancinelli family.)

Windsor, Ontario, Canada, is located across the Detroit River. Here, members of Engine Company No. 2 stand by for orders to respond to another fire in the areas of unrest. Detroit and Windsor have provided international mutual aid to each other when called upon—not too many other cities can boast that fact. (Courtesy of the Joe Mancinelli family.)

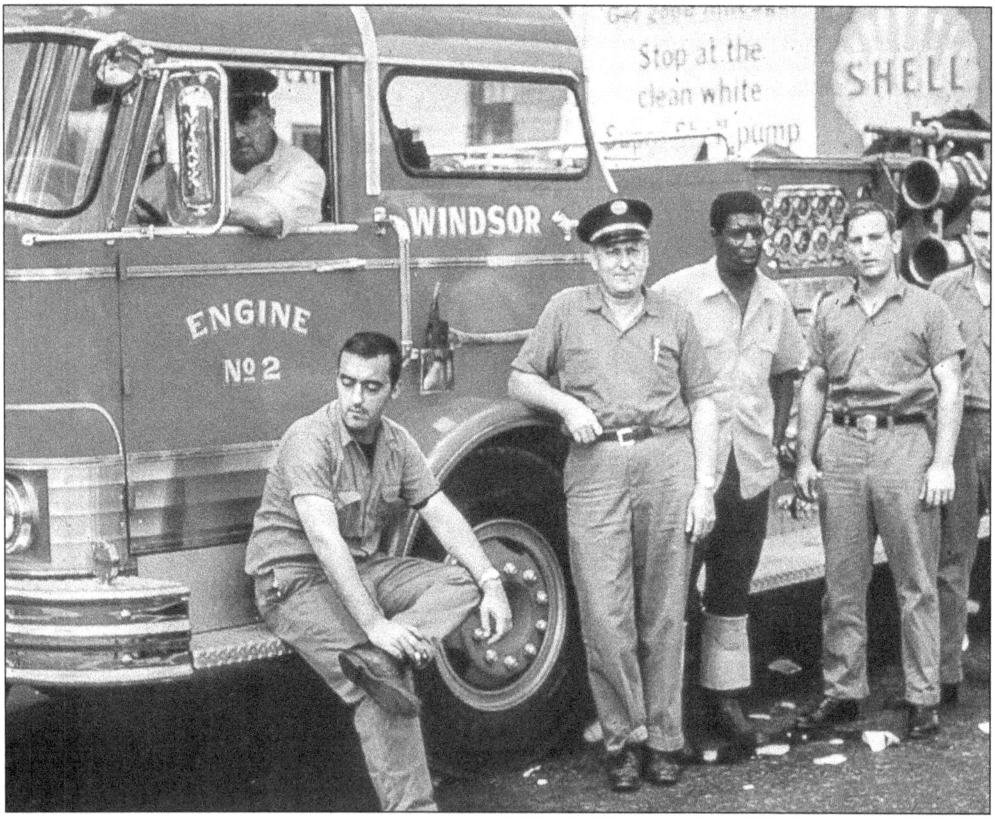

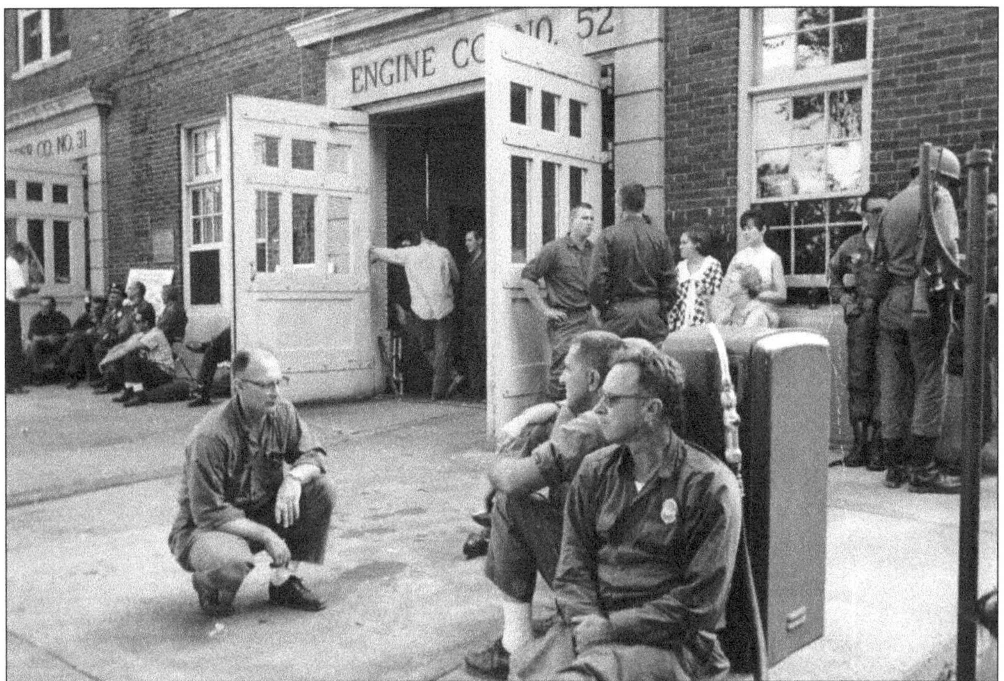

The firehouse of Engine Company No. 52 and Ladder Company No. 31 served as a rehab area for firemen to rest, eat, bathe, tend to injuries, and visit with family members who ventured out to see their loved ones. (Courtesy of the Detroit Firemen's Fund.)

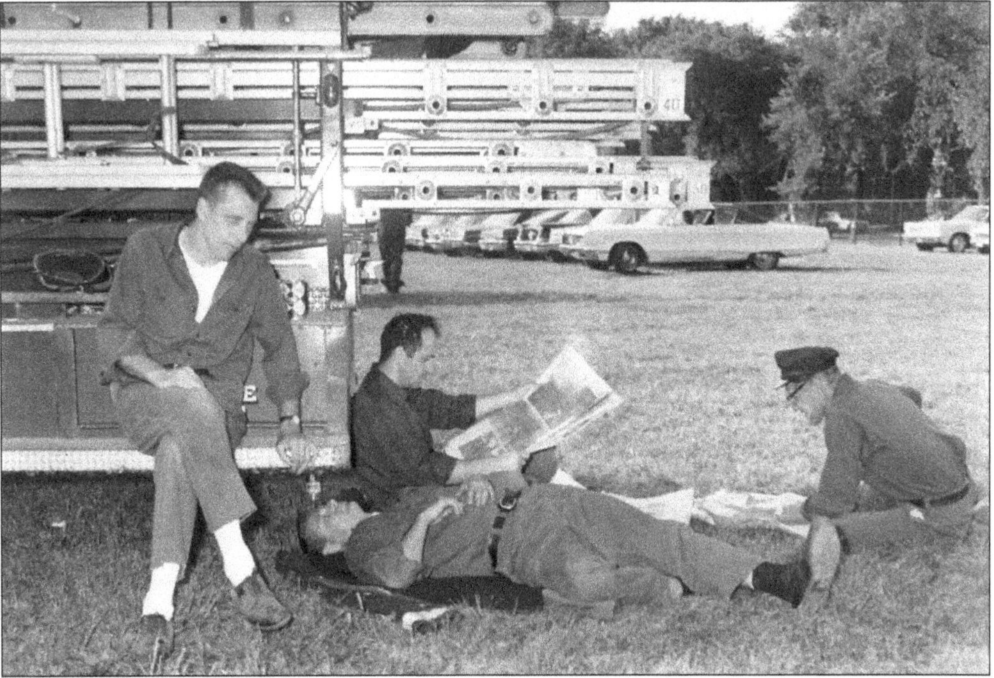

Firemen relocated to a high-security area to take time to rest and refresh between fires. Those coming from home would be assigned to fire companies that needed relief before reaching their assigned company. (Courtesy of the Detroit Firemen's Fund.)

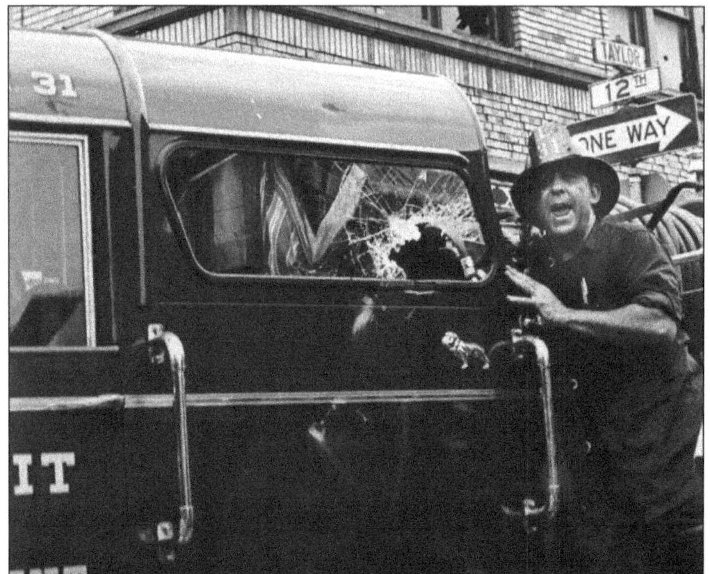

As the riots went on and firemen and their vehicles had to work in areas under intense conditions, damage still continued to occur to both the rigs and the members. Here, a fireman from Engine Company No. 31 calls for help as he and the rig are under attack from items being thrown at them. (Courtesy of the Detroit Firemen's Fund.)

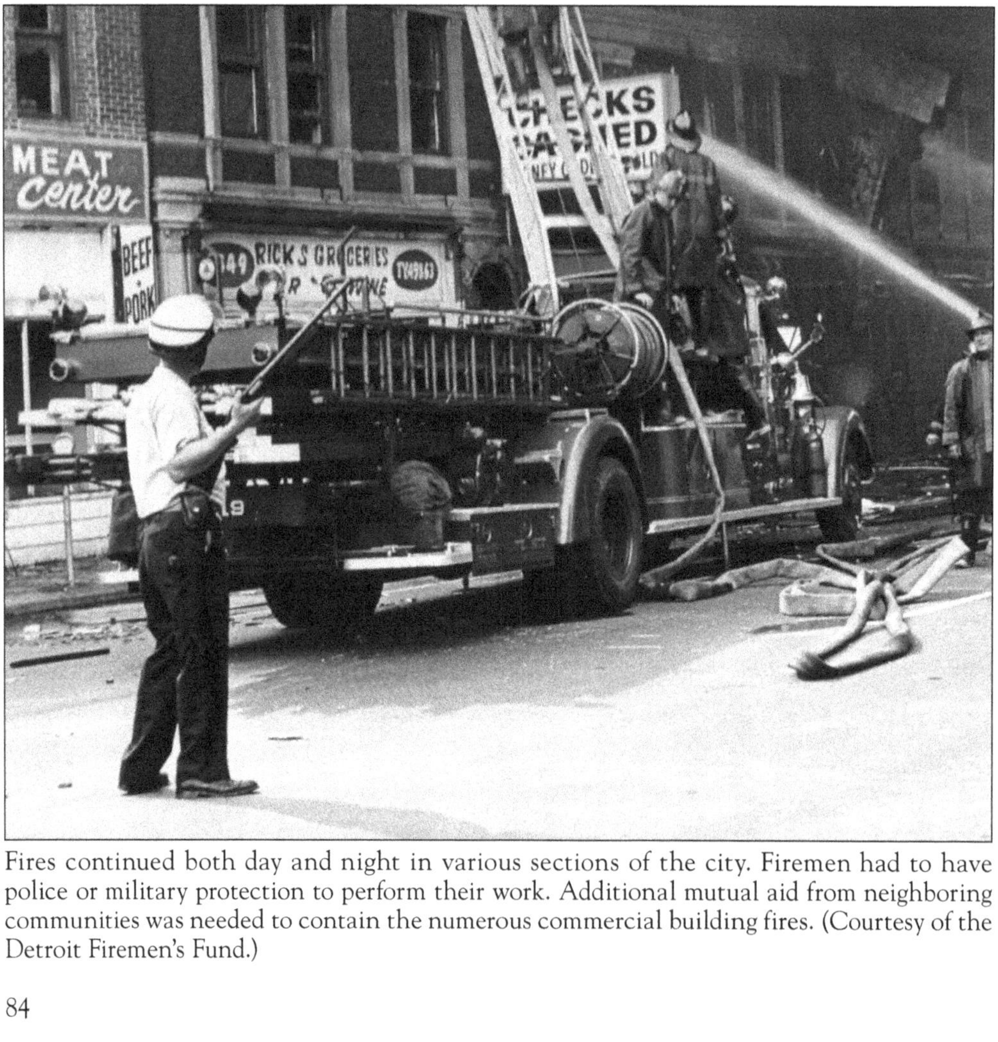

Fires continued both day and night in various sections of the city. Firemen had to have police or military protection to perform their work. Additional mutual aid from neighboring communities was needed to contain the numerous commercial building fires. (Courtesy of the Detroit Firemen's Fund.)

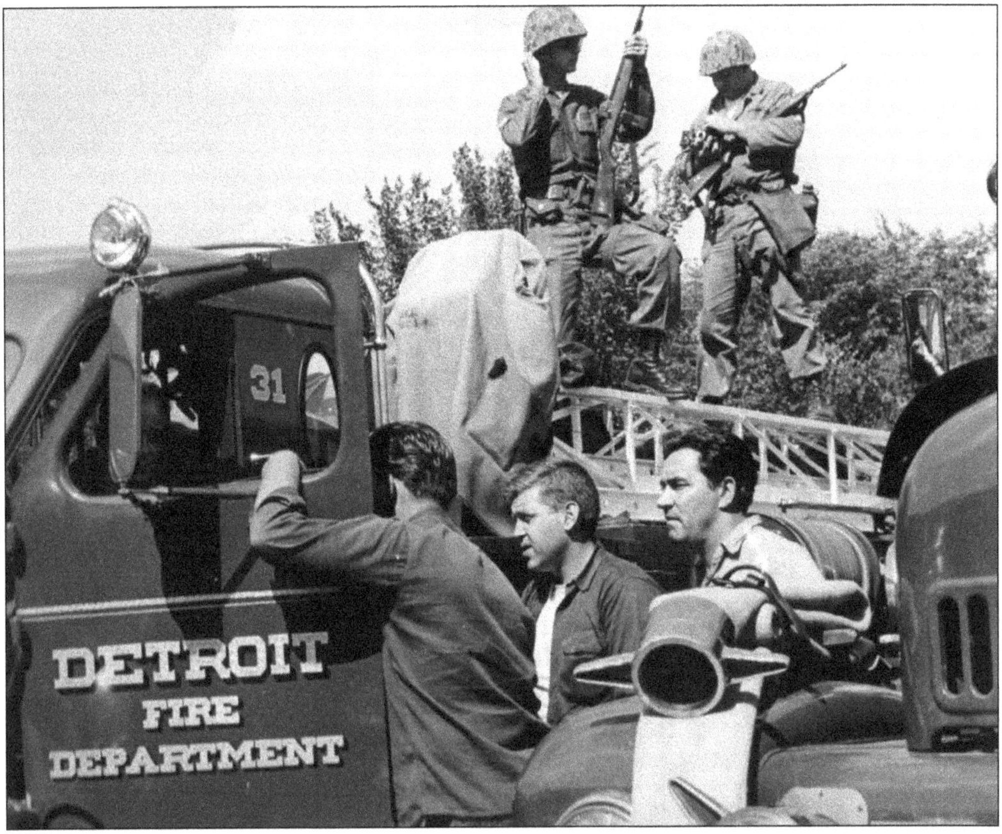

Firemen took advantage of breaks when there was a lull in the action or when additional firemen from other communities relieved them. Here, members of Ladder Company No. 31 check their equipment before venturing out to another fire while soldiers standing on the aerial ladder do the same. (Courtesy of the Detroit Firemen's Fund.)

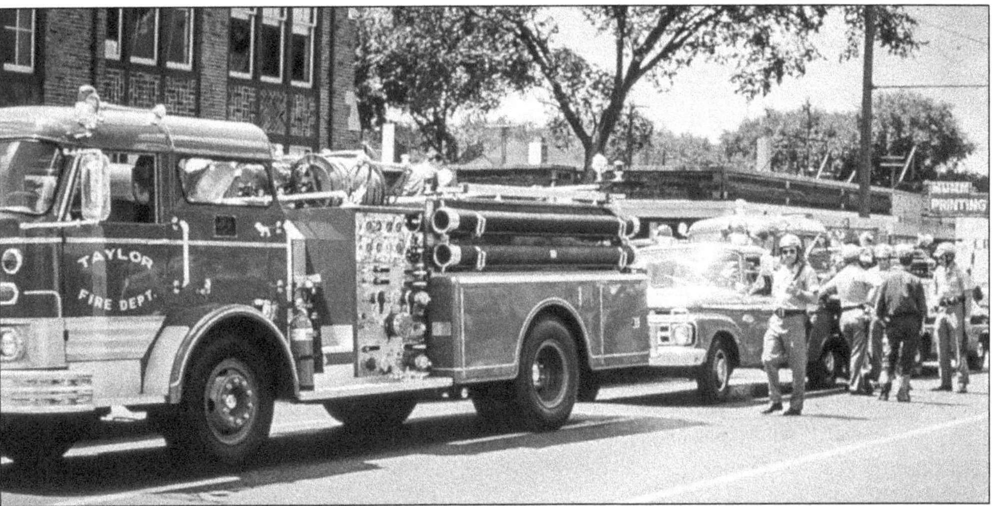

Members of the Taylor Fire Department arrive in the staging area under high-security conditions, with police heavily armed, before being sent out to help their firefighting brothers. (Courtesy of the Joe Mancinelli family.)

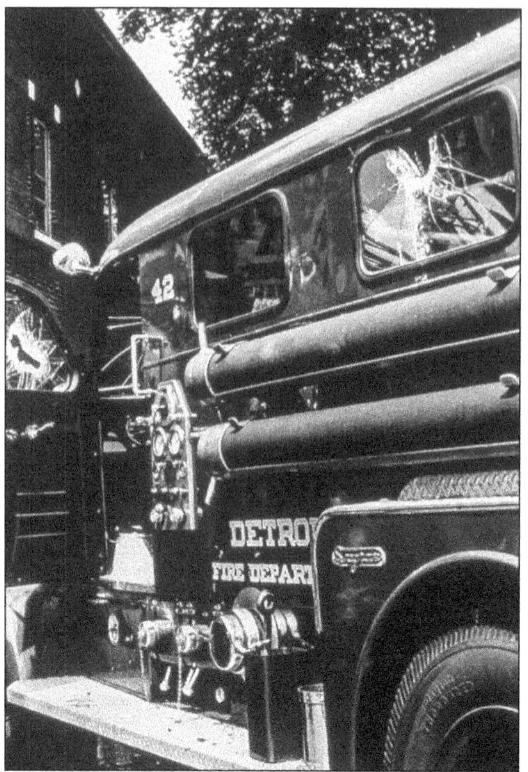

Damage to fire department vehicles continued. Here, Engine Company No. 42 has been hit with rocks, bottles, bricks, and pipes as they drove to this fire. When possible, the repair shops would make major repairs, but most of the time, the rigs had to keep their damaged parts as a badge of courage. (Courtesy of the Joe Mancinelli family.)

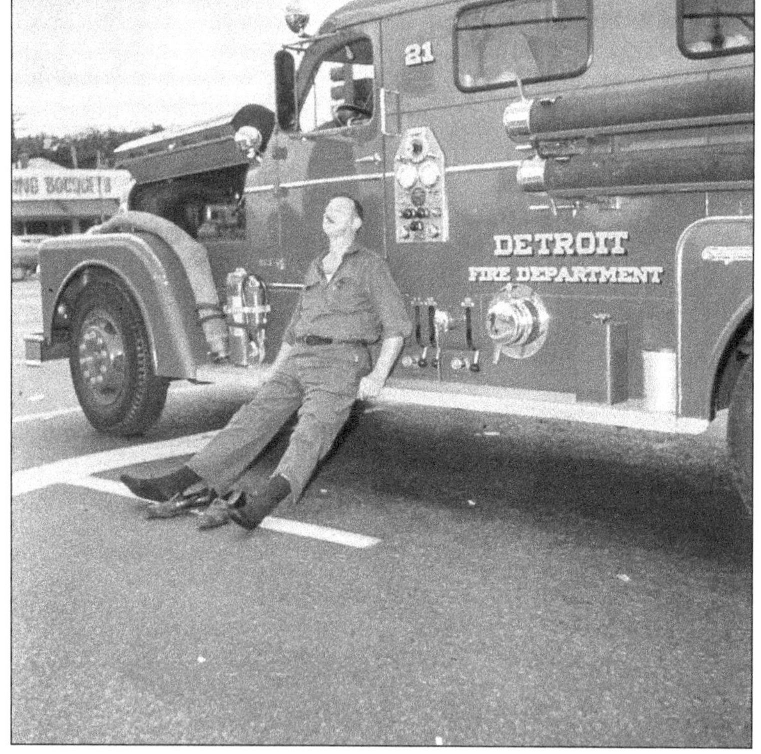

The driver for Engine Company No. 21 takes a well-deserved smoke and nap break after having worked at several fire scenes; many during this period would have been extra-alarm fires under normal conditions. (Courtesy of the Joe Mancinelli family.)

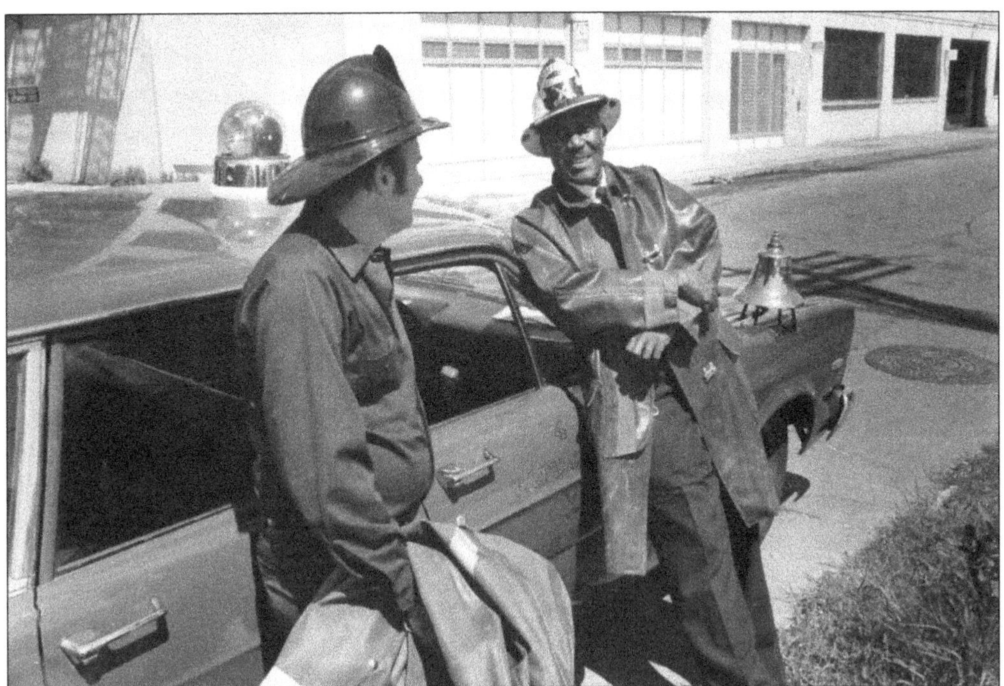

Detroit's 4th Battalion chief Marlene Taylor and his aide discuss matters at hand while operating at the 1967 Riots. Over the five-day period, chiefs responded with fire companies to fire scenes. Chief Taylor was revered by many members of the DFD. (Courtesy of Matt Lee.)

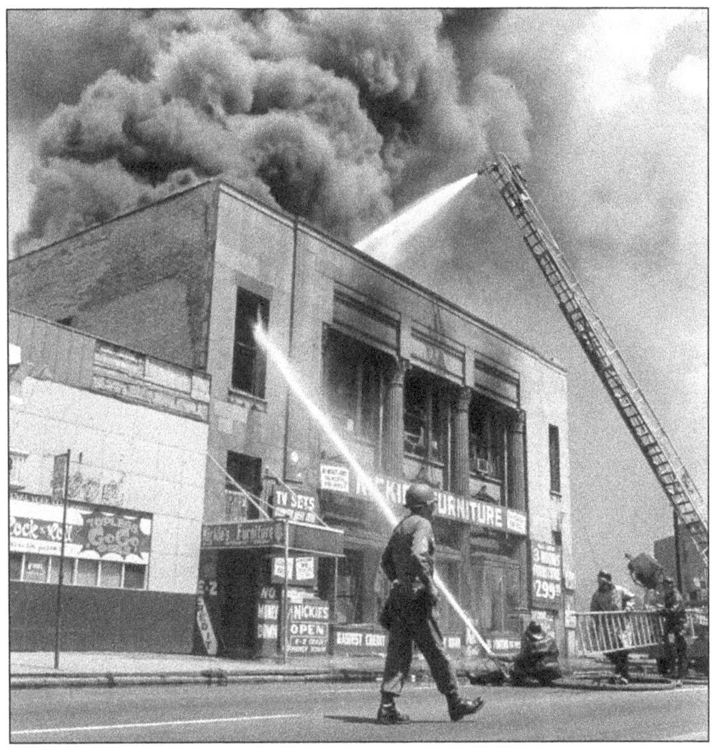

Here, at one of the last fires of the riots, soldiers walk the perimeter of a commercial building. The military had a job to provide safety to Detroit firemen and police officers. Nineteen National Guardsmen were injured during the riots; there was no report of federal soldiers injured. (Courtesy of the Joe Mancinelli family.)

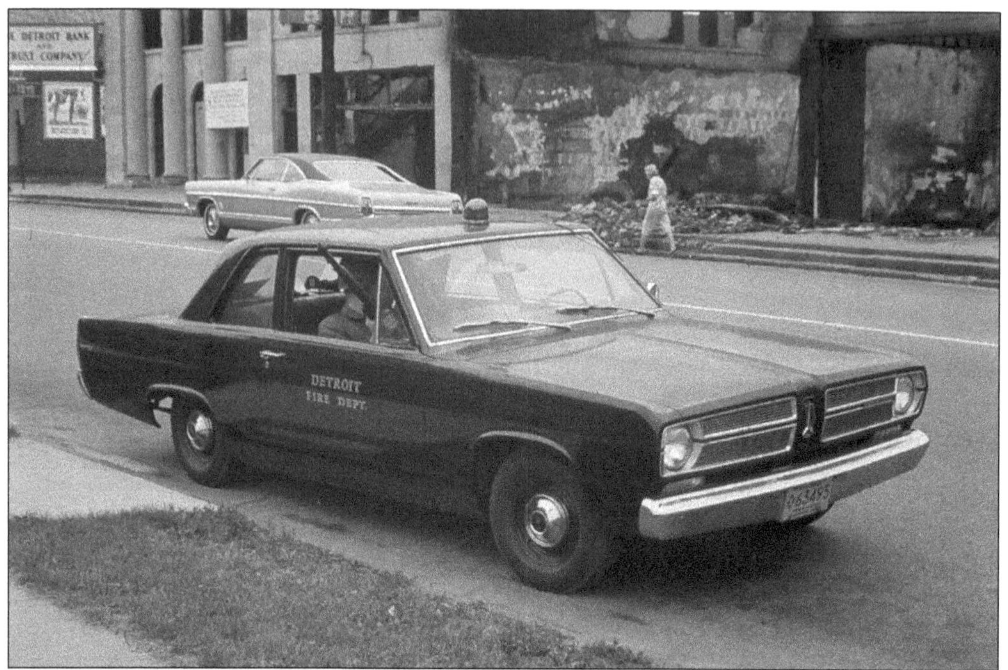

Detroit firemen who worked during the riots of 1967 will always be able to remember the incidents that they responded to. A chapter in the nation's history was written over that five-day period. Here, a fire department vehicle is being used by National Guard soldiers with weapons to respond to locations where needed. A volunteer at the Salvation Army Canteen Unit, who fed firemen, stated, "The valiant firefighters [did] a magnificent job under terrifying conditions." (Courtesy of the Detroit Firemen's Fund.)

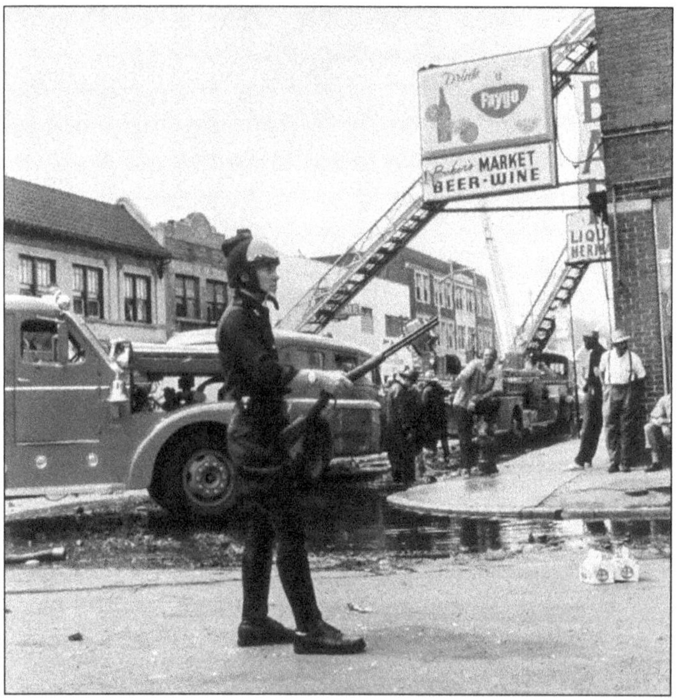

As things started to return to normal, images captured of the riots were like permanent wounds for many residents, public safety officers, and public officials, as the healing process really never happened. Many residents packed up and moved out of Detroit, never to return. The declining population continued for many years, and the fires continued to increase with the rise of vacant properties. (Courtesy of the Detroit Firemen's Fund.)

Six

1970s–2000

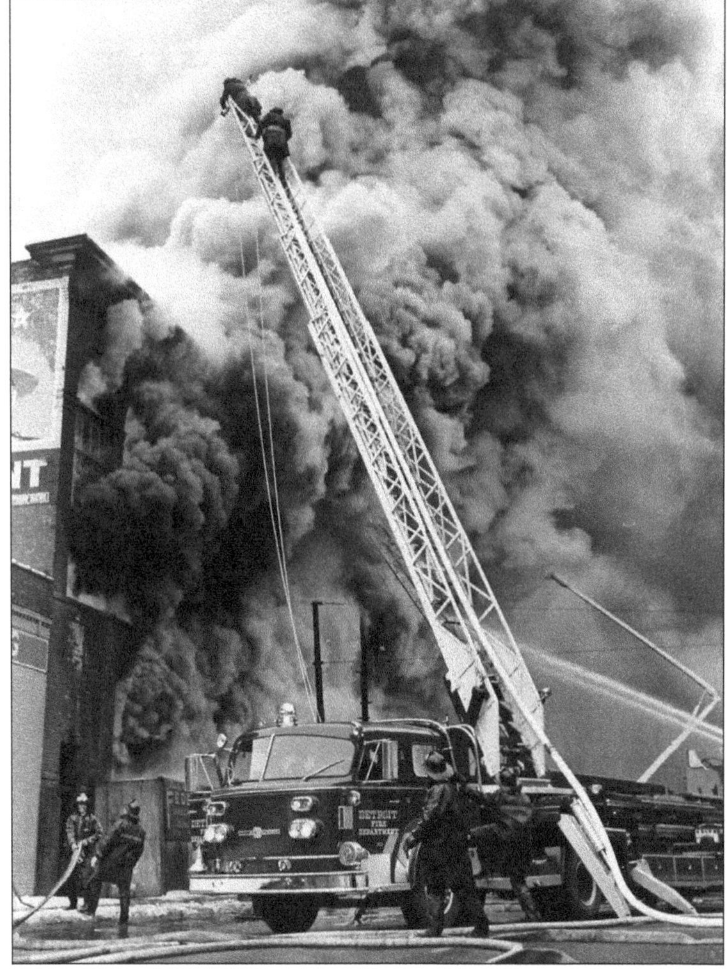

In subzero temperatures, Ladder Company No. 1 is setting up the aerial pie to go to work at a multiple-alarm fire at West Fort and Eighth Streets. The fire was in a former hotel about to be torn down in January 1970. Note the guide ropes connected to the aerial pipe; this would be operated by a fireman on the ground, in this case behind the cab. (Courtesy of Matt Lee.)

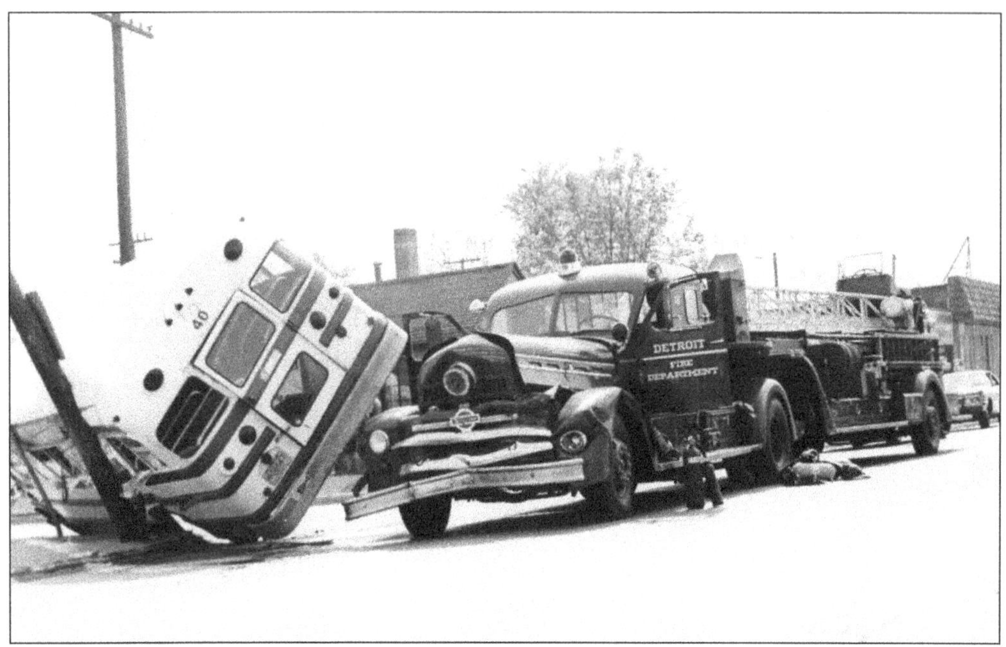

Ladder Company No. 18 had a 1953 Seagrave aerial truck that was involved in a collision with a private school bus on East Seven Mile Road. The school bus yielded for a passing engine but failed to yield for the oncoming ladder truck. The accident happened in May 1972, and the incident prompted the department to explore installing air horns on the ladder trucks that were housed with other vehicles. (Courtesy of Matt Lee.)

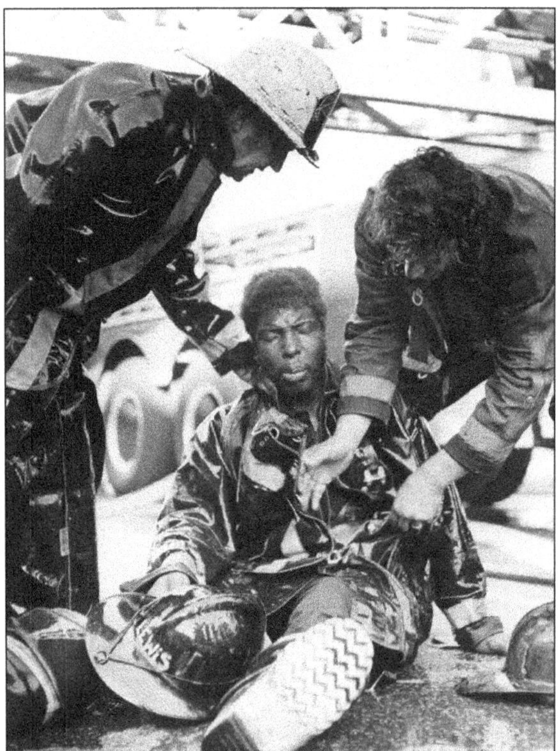

This Detroit fireman was overcome by smoke after fighting a fire in August 1974. Upon exiting the building, he collapsed and was attended to by a fellow firefighter and a battalion chief while working at the third-alarm fire in a super market located at 10520 Mack Avenue. (Courtesy of Matt Lee.)

"Picking Up" Pipeman Premo Damiani of Tactical Mobile Squad No. 1 and other firemen assist in loading two-and-a-half-inch hose back onto the Seagrave sedan pumper of Engine Company No. 8 during a 1974 east-side multiple alarm. Fireman Damiani advanced in his career and later became chief of the department. (Courtesy of Matt Lee.)

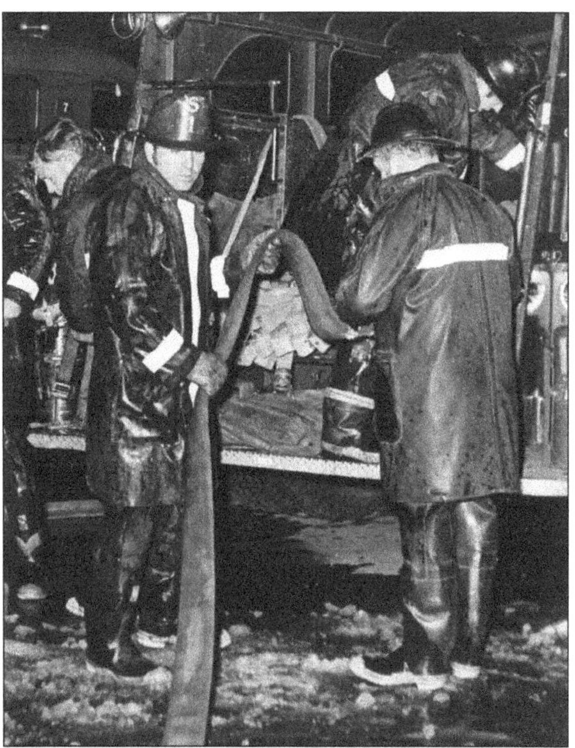

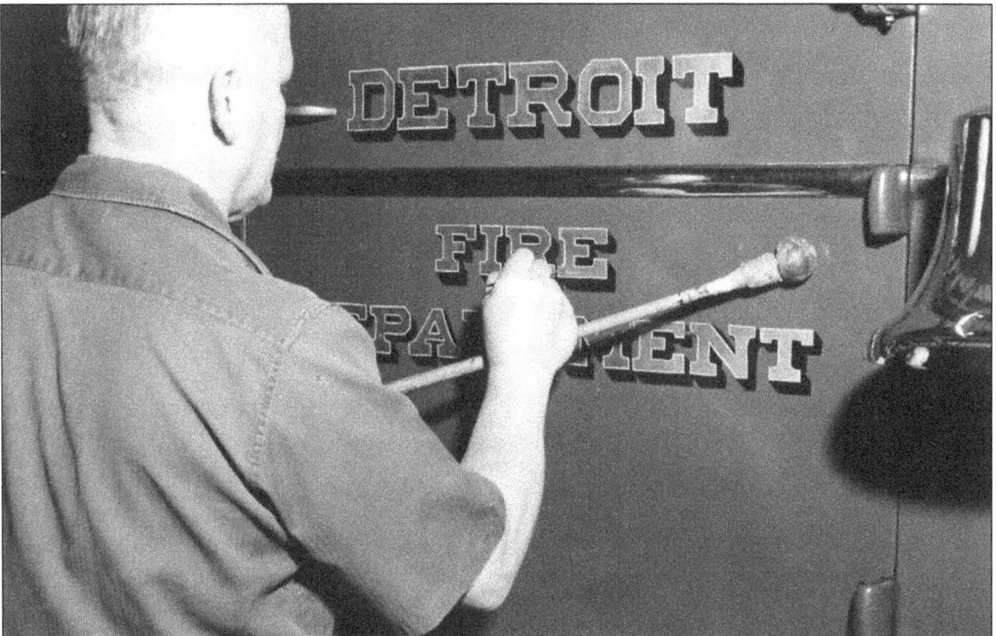

The famous logo found on all Detroit Fire Department vehicles was hand-painted by a member of the DFD Repair Shops. A steady hand always made sure that the name appeared with pride on every fire department vehicle on the streets of Detroit. This member of the Repair Shops was also responsible for painting all the door signs at Detroit Fire Department Headquarters. (Courtesy of Matt Lee.)

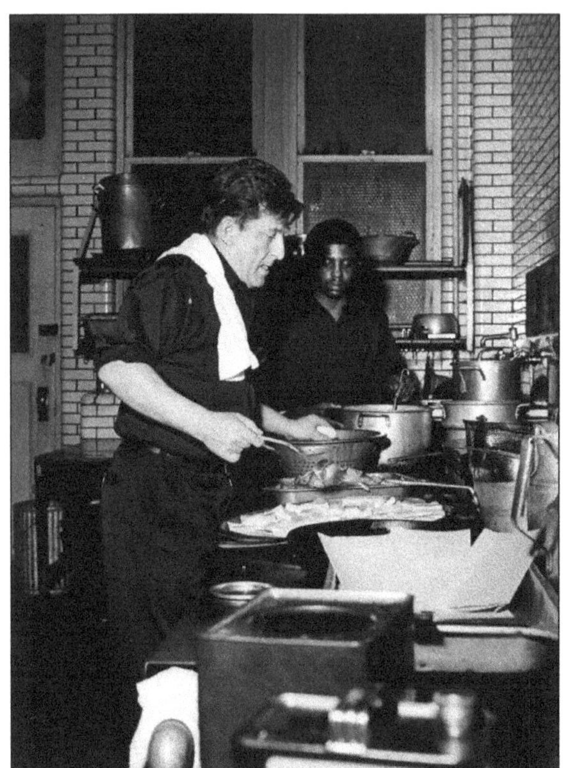

The firefighters of the Detroit Fire Department always take pride in preparing the meals of the day at firehouses. Here, two firemen at the headquarters firehouse work on getting the meal prepared before the next run would sound. The subway tile is a standard backdrop in all DFD firehouses. The wear and tear on the pots and pans is also a standard due to constantly being used. (Courtesy of Matt Lee.)

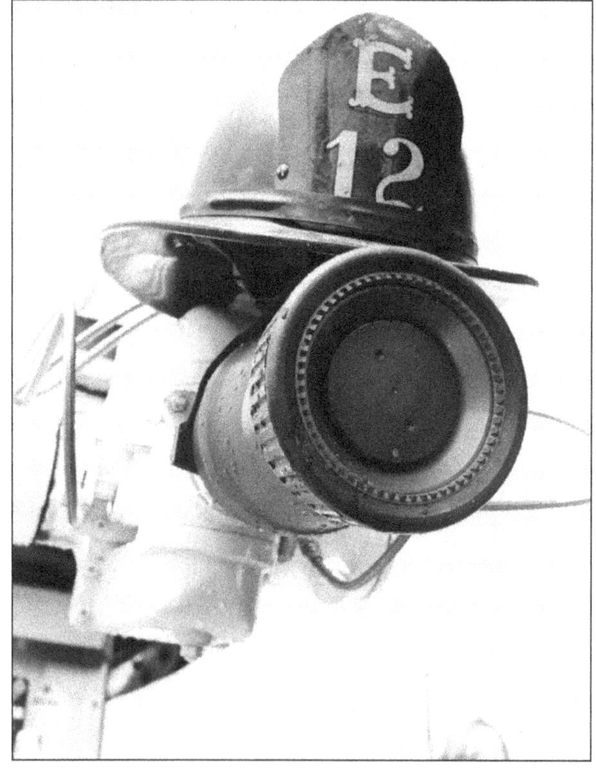

Here, a helmet from Engine Company No. 12 has been placed on an electric-operated nozzle from a tele squirt. The famous shields found on most Detroit firefighters' helmets were hand-painted in the repair shops. Today, those shields are still being made by two Detroit firefighters who create each with care and pride. (Courtesy of Matt Lee.)

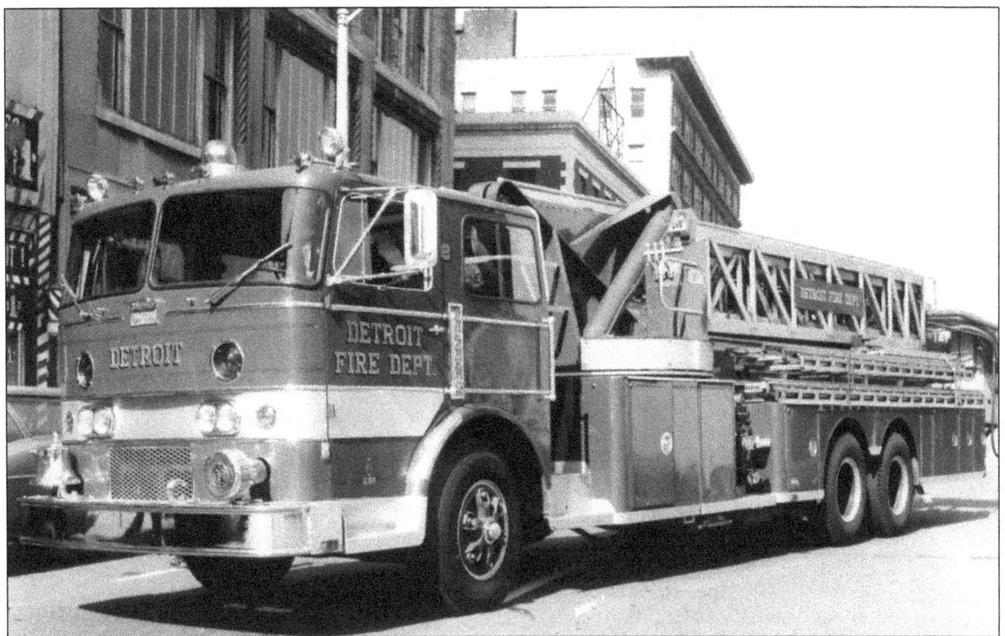

In 1970, Ladder Company No. 6 operated this Sutphen three-section aerial tower. The unique feature of the tower was the rear platform (bucket) that operated two pre-piped aerial nozzles, which allowed double the water versus the standard aerial pipe. This was the last unit delivered in the color red. (Courtesy of Matt Lee.)

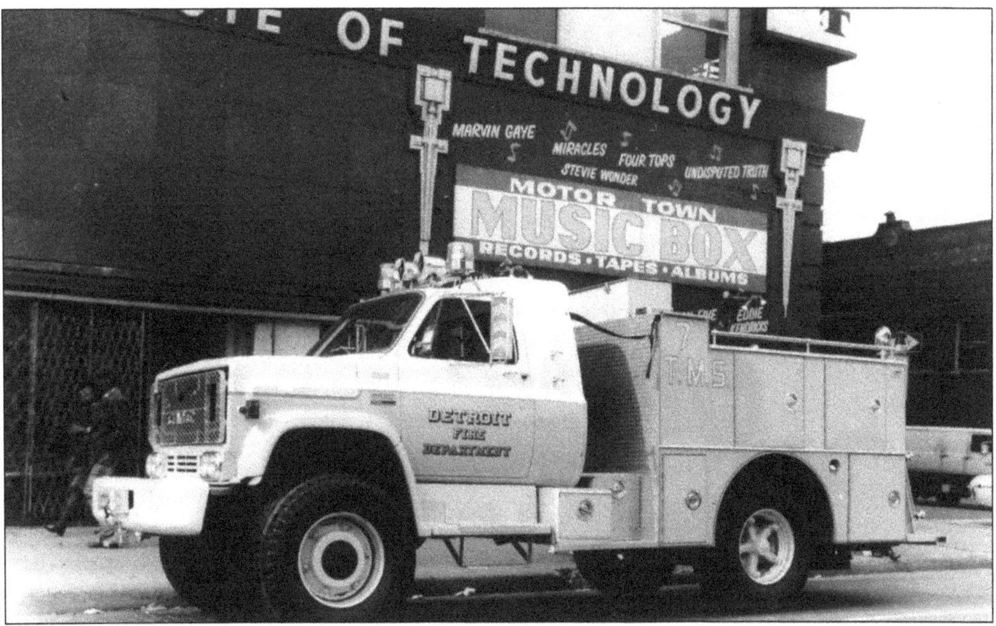

Tactical Mobile Squad (TMS) No. 7 ran with this unusual rig using a lime yellow and white color scheme. The 1973 GMC had a commercial chassis with a walk-through section located behind the cab. TMS No. 7 ran out of Greenfield Road and Fenkell Avenue, the quarters of Engine Company No. 53, in northwest Detroit. The vehicle was delivered for the expressway (ditch) steep embankments and off-road use. (Courtesy of Matt Lee.)

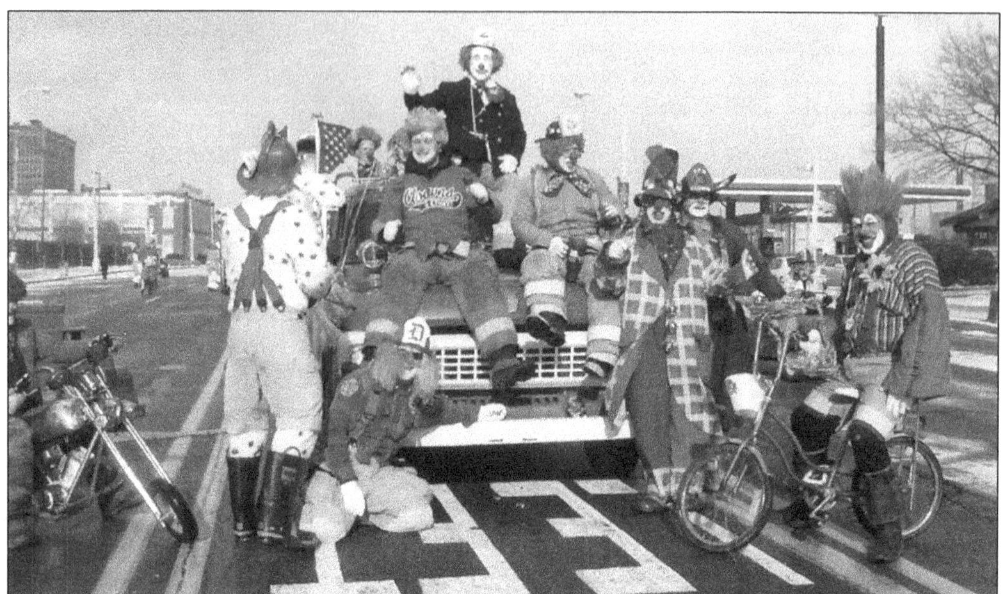

The Detroit Fire Department Clown Team was organized to entertain the public at the annual field-day event, parades, and burn camp visits to name a few events. The clown team has changed clowns over the years, but is still in existence today. Firemen who joined the team had a name for the clown they performed as and had a makeup scheme so that they looked the same every time they appeared in costume. (Courtesy of Bill Eisner.)

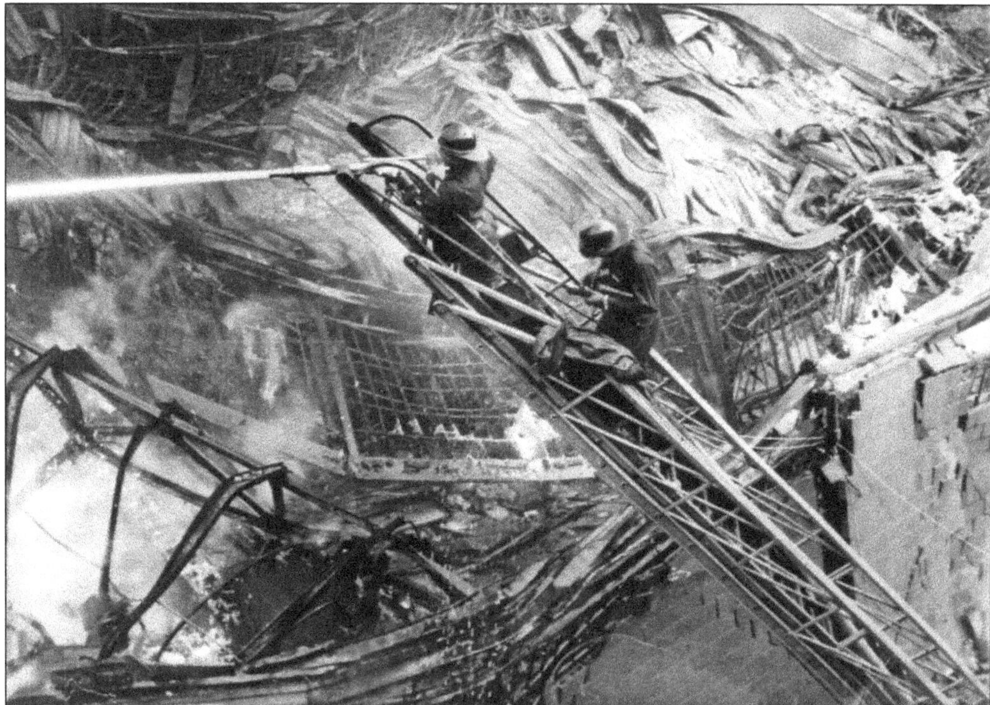

This is a rare aerial view of two Detroit firemen directing the stream of their extended ladder pipe on hot spots at a multiple-alarm fire. The building collapsed under the high temperatures, which caused the building's metal roof to twist and fall. (Courtesy of Matt Lee.)

This award-winning photograph, "What Is a Fireman" by Detroit's Capt. Joseph Mancinelli, shows two DFD firemen from Engine Company No. 8 working a hose line at a multiple-alarm fire in ice and snow and subzero temperatures. Note the three-quarter rubber boots and rubber fire coats. (Courtesy of the Joe Mancinelli family.)

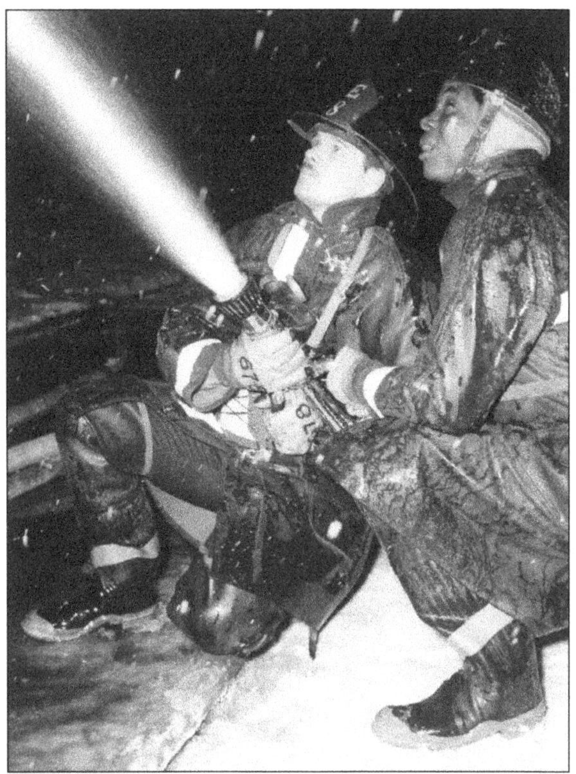

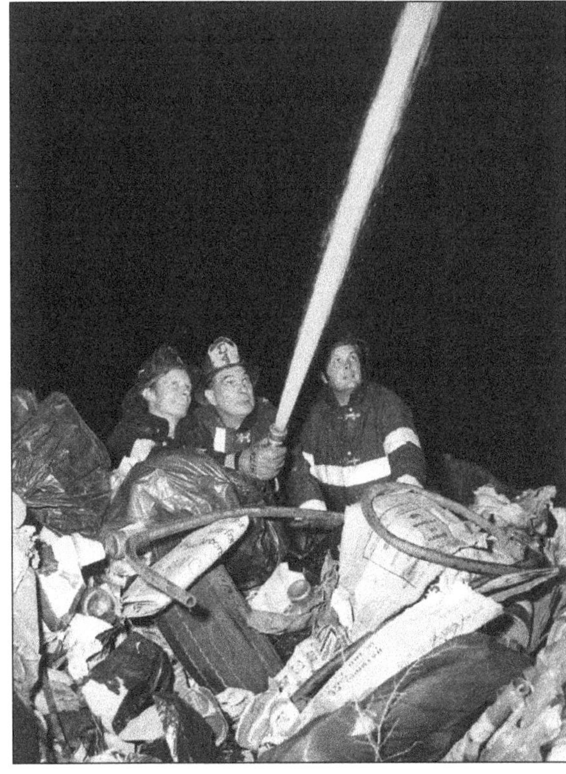

Members of Tactical Mobile Squad No. 4 operate a hose line at a fire scene. Note the firemen are working on a large pile of garbage to achieve their goal to get access to hit the fire from the outside at that vantage point. (Courtesy of Matt Lee.)

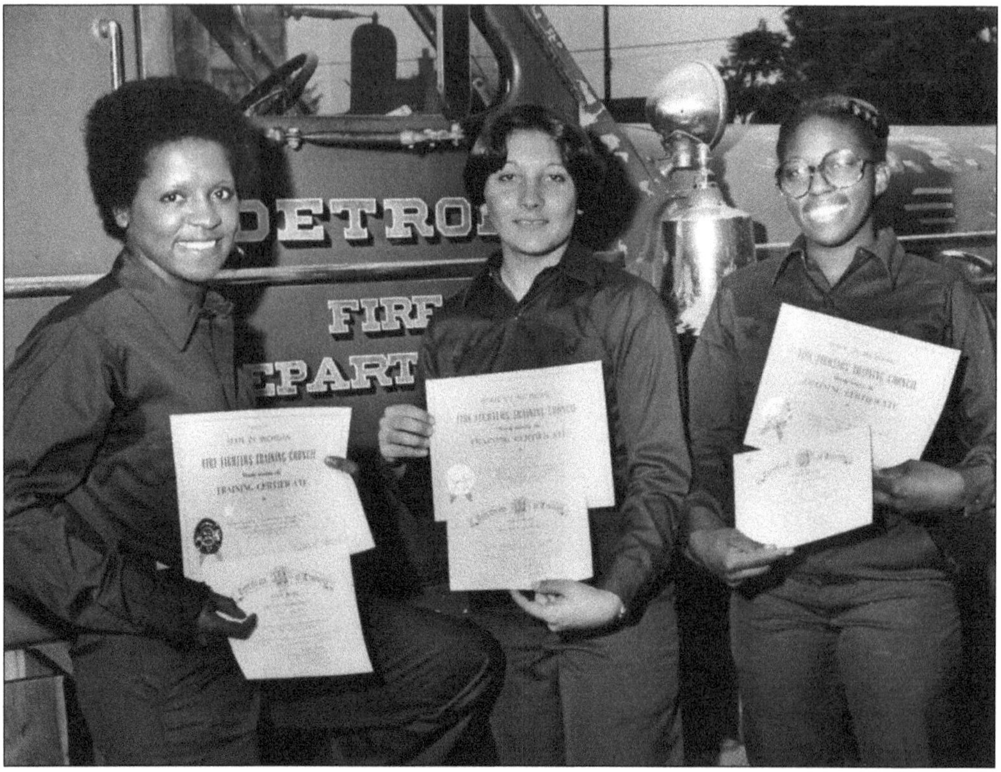

The first women to qualify as Detroit firefighters are, from left to right, Harriet Saunders, Sandy Kupper, and Theresa Smith; all graduated from the DFD Training Academy in September 1977. The women were assigned to engine companies to begin a four-month probationary tour of duty. (Courtesy of Matt Lee.)

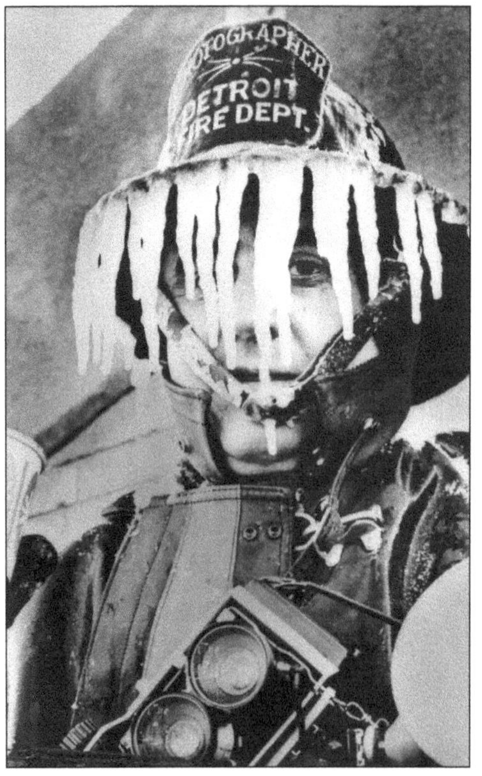

Capt. Joseph Mancinelli, the Arson Squad photographer, takes time to pose for a photograph with his ice-covered helmet at the scene of a multiple-alarm fire in below-zero conditions. (Courtesy of the Joe Mancinelli family.)

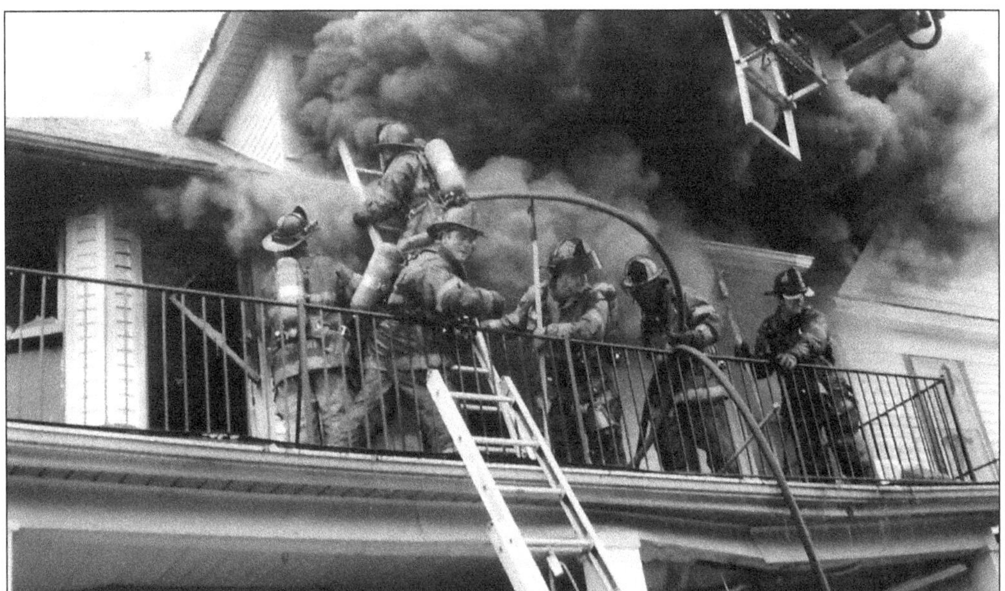

An annual event in which Detroit received worldwide coverage was "Devil's Night" on October 30, the night before Halloween. It dates back to the 1930s, but back then it did not involve acts of arson. Here, firefighters work aggressively to bring this Devil's Night blaze under control. (Courtesy of Bill Eisner.)

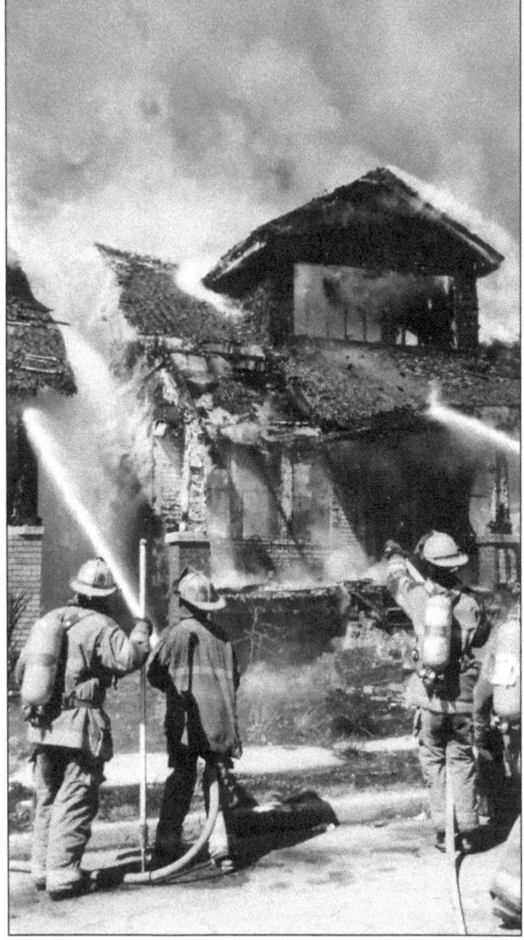

In the 1990s, community response prompted a change in the activity name to "Angels' Night." The number of fires started to decreased. In 1984, a record number of fires took place over Devil's Night—more than 800 fires were recorded. (Courtesy of Bill Eisner.)

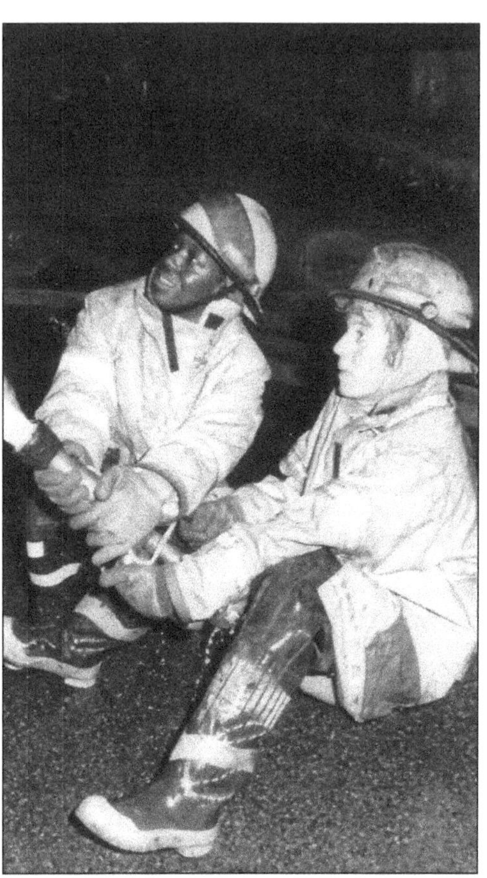

Here, two Detroit female firefighters, Joyce Zmudczynski and Verdine Pierce, direct a stream of water into a building at a multiple-alarm fire. (Courtesy of Chief Joyce Stoll.)

Detroit fire sergeant Stephen Babicz of Ladder Company No. 19 carries a child to a waiting ambulance in an effort to save his life from a house fire in which seven children died. This fire was located at Mack Avenue and Chene Street. The photograph was nominated for a Pulitzer Prize. (Courtesy of Bill Eisner.)

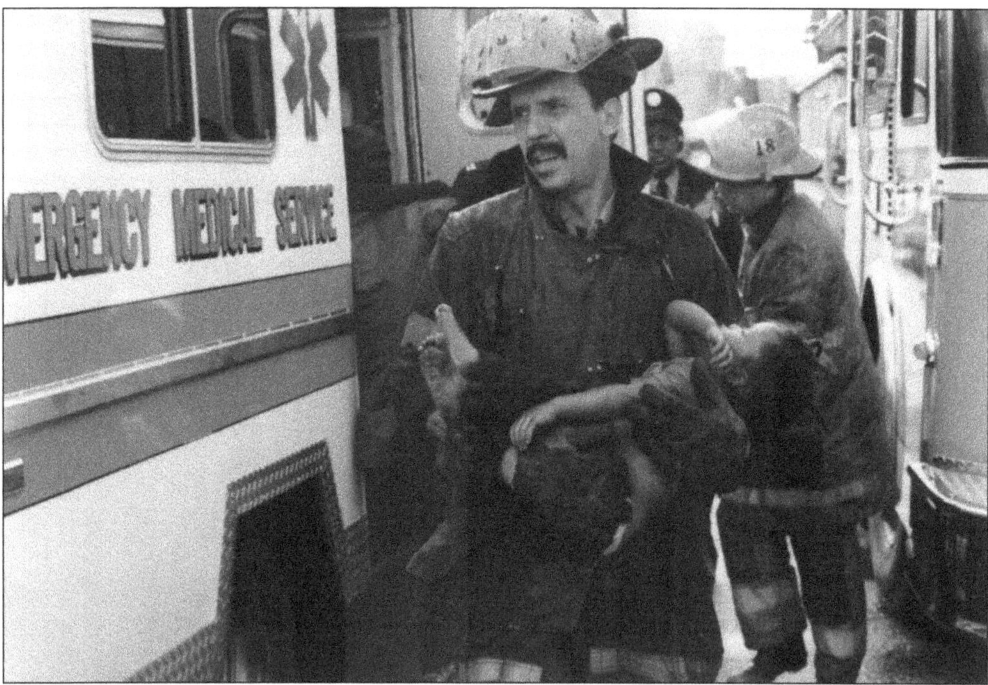

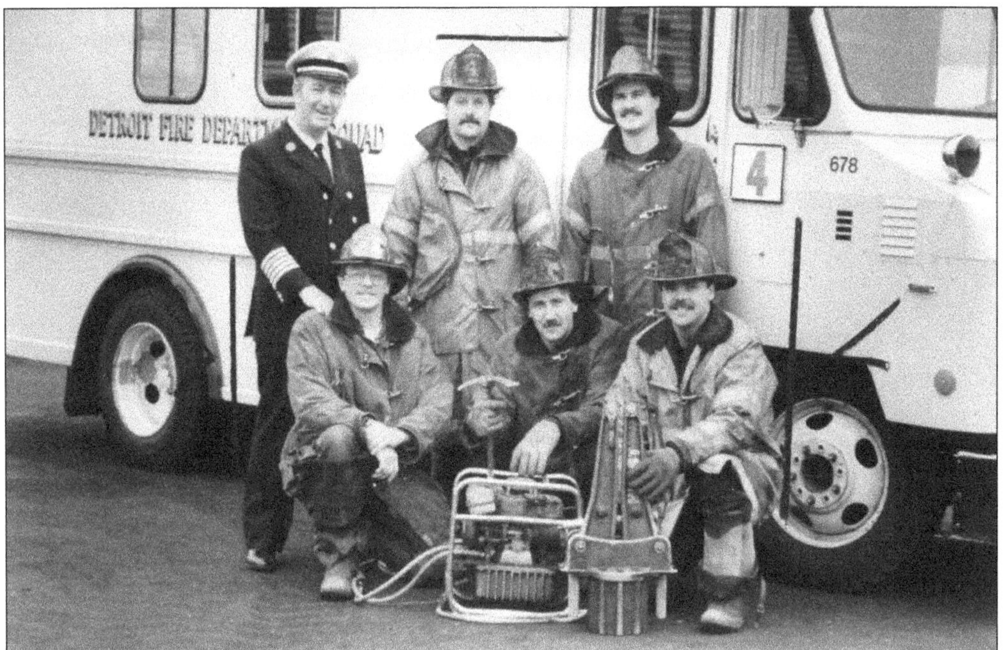

In November 1985, members of TMS No. 4 pose with their new addition to the squad, the Hurst rescue tool. They display the portable motor and the JAWS, or spreaders. This tool was used for car accidents and forcible entry at commercial buildings due to tightened security. (Courtesy of Jon Soave.)

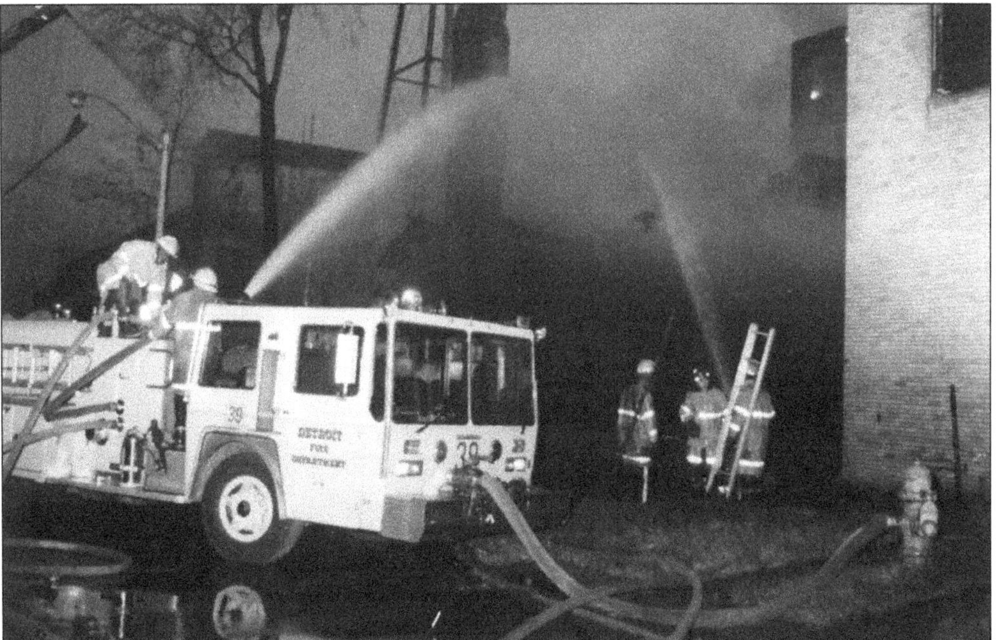

A generation of fire-vehicle change took place in 1970s with the color change from red to lime-yellow. The change was designed to make the vehicle more visible based on national studies across the country for firefighter safety. Here, Engine Company No. 39 operates at a box alarm fire in a commercial building. (Courtesy of Matt Lee.)

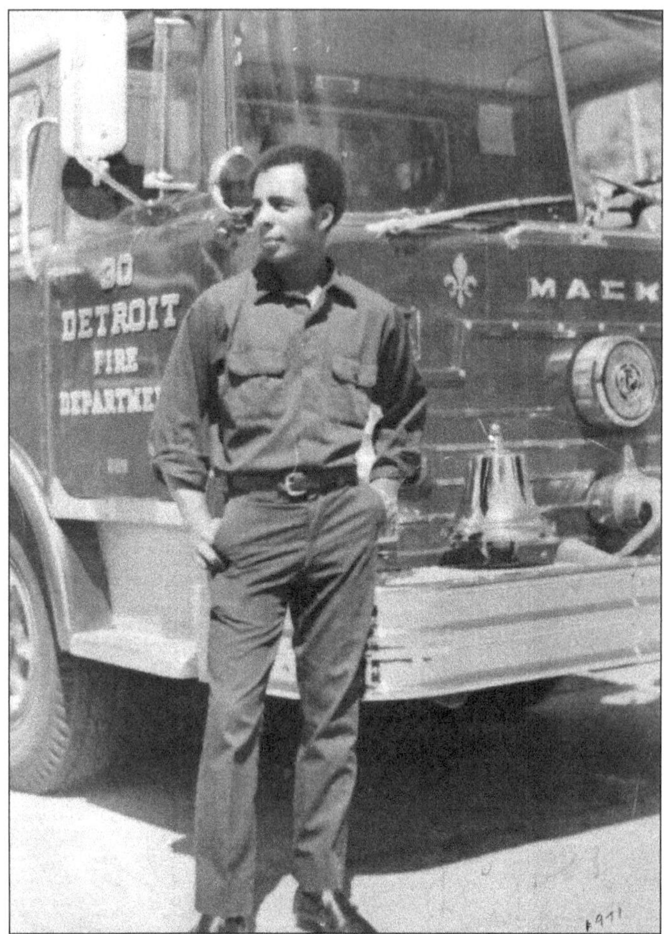

Detroit fireman Rodney Parnell from Engine Company No. 30 poses alongside his rig. Engine Company No. 30 is currently located in northwest Detroit. Today, Chief Parnell oversees community relations and fire safety. (Courtesy of Matt Lee.)

Firefighters from fire companies in southwest Detroit direct hand lines into a building across the street at Witt and Lawndale Streets. The house fire was an arson attack. It was so intense that firefighters were not allowed to enter the structure for fear of the building collapsing. (Courtesy of Bill Grimshaw.)

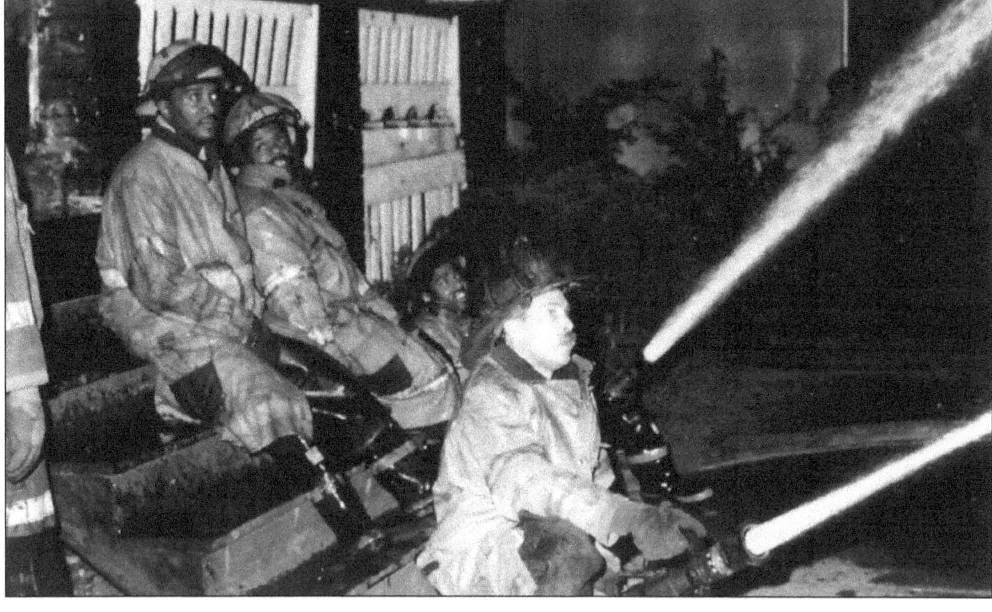

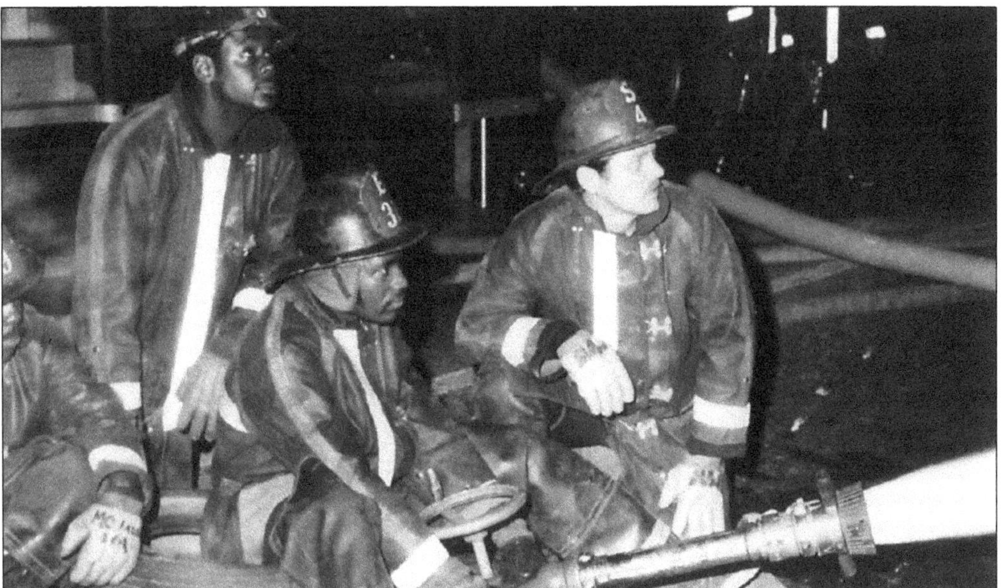

Firefighters from Engine Company No. 31 and Squad Company No. 4 direct the stream from a deck gun at this multiple-alarm fire located at Joy Road and Dexter Street into a commercial building. Deck guns help provide firefighters with a large volume of water, flowing hundreds of gallons of per minute with pressures that help penetrate deep into the buildings to spots firefighters cannot enter due to safety concerns. (Courtesy of Bill Grimshaw.)

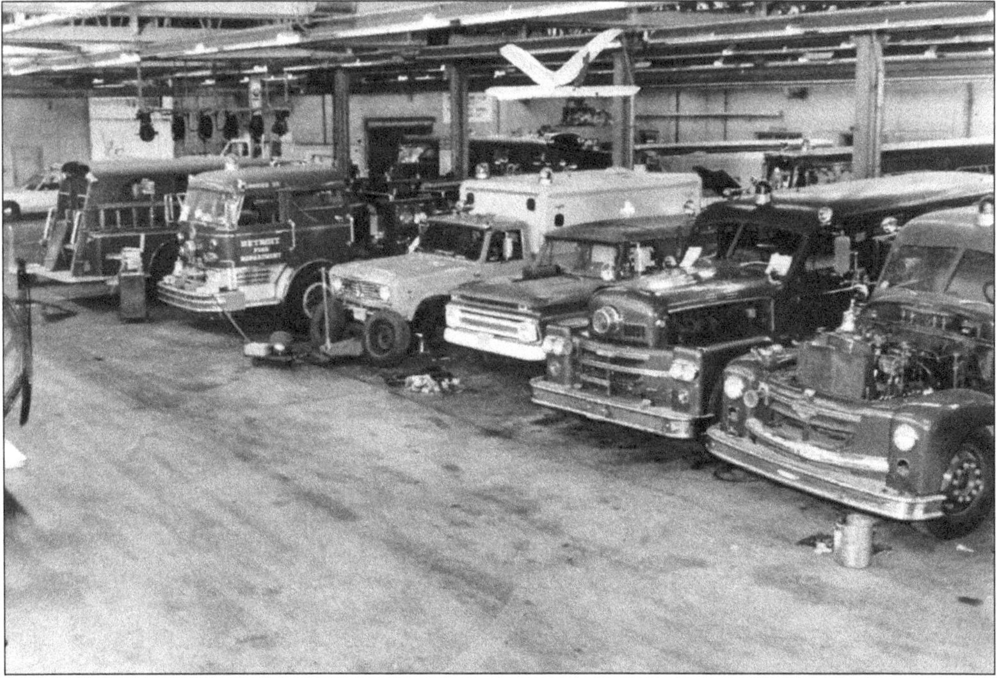

The repair shops have not changed over the years and nor has the wait for vehicles in need of repair. Here, vehicles in the shop wait either for parts or funding to be repaired. (Courtesy of Dan Jasina.)

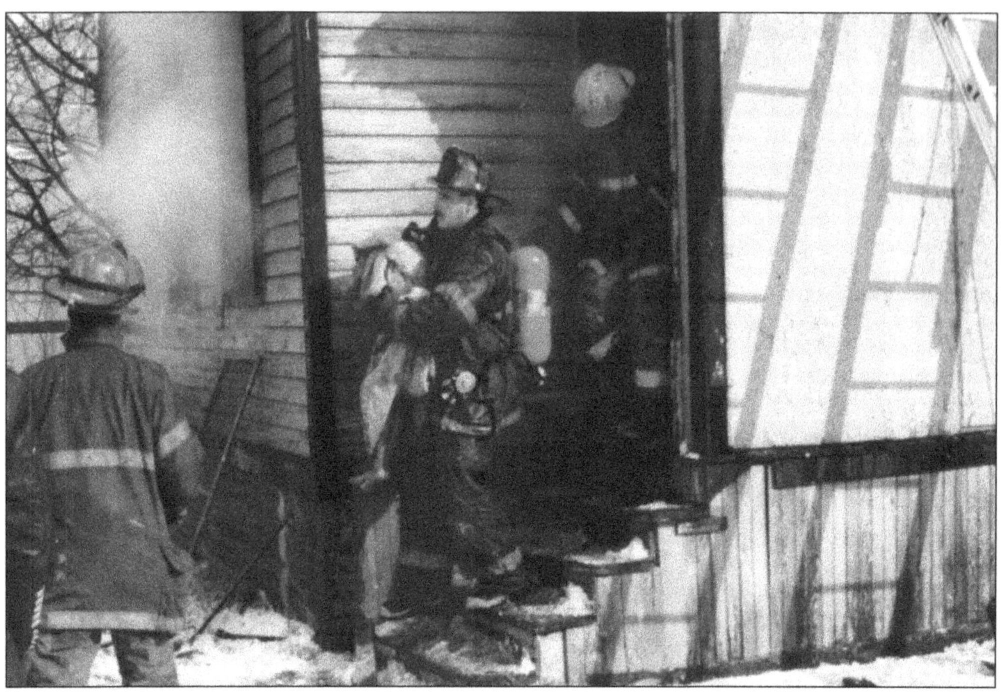

Detroit Firefighter John Cowan of Engine Company No. 23 carries out a child to a waiting ambulance from a house fire. Sadly, seven children died in this house fire. (Courtesy of Bill Eisner.)

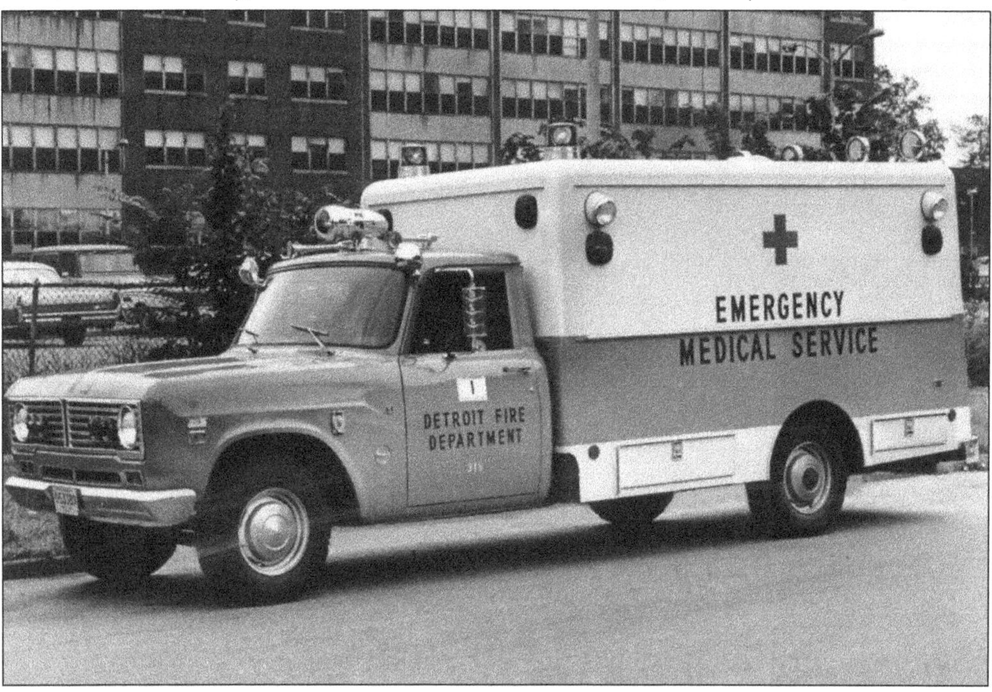

In the 1970s, emergency medical services changed with the advent of emergency medical technicians (EMT) and paramedics. Here, a new ambulance has been assigned to Detroit's new EMS program. EMTs and paramedics were hired to provide a new service to residents. (Courtesy of the Detroit Firemen's Fund.)

On June 11, 1982, a man entered the Buhl Building in Detroit, Michigan. His attorney was handling an insurance matter for him, and the man wanted his check. It was for $2,500, and the man intended to get that money or seek vengeance. He rode the elevator to the eighth floor. When he entered the lawyer's office, he was carrying a sawed-off shotgun and a gasoline fire bomb. Whatever occurred from that point on is uncertain. The man shot a woman and threw the fire bomb against the office wall. The ensuing explosion and fire scattered the occupants of the eighth floor. Many ran to windows, cut off by the raging fire, while some were caught in smoke-filled hallways. The fire alarms went off, and firefighters arrived on the scene. Panic gripped the eighth floor of the Buhl Building. (Courtesy of Matt Lee.)

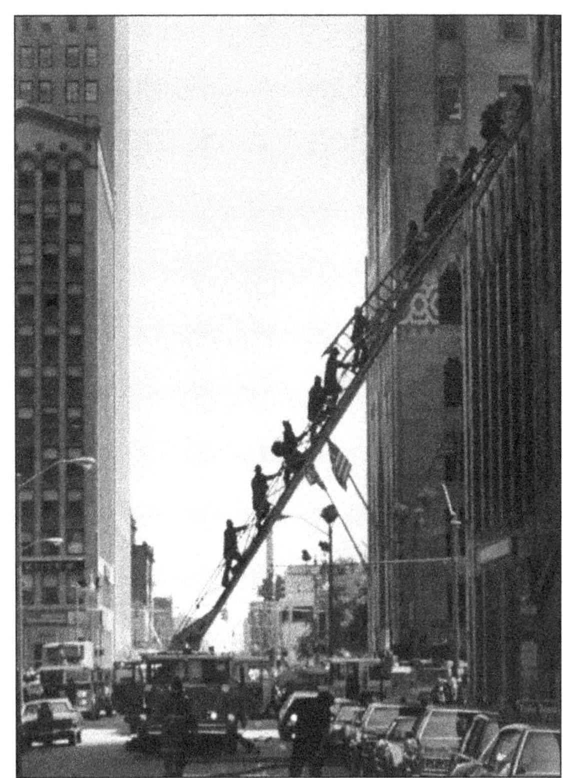

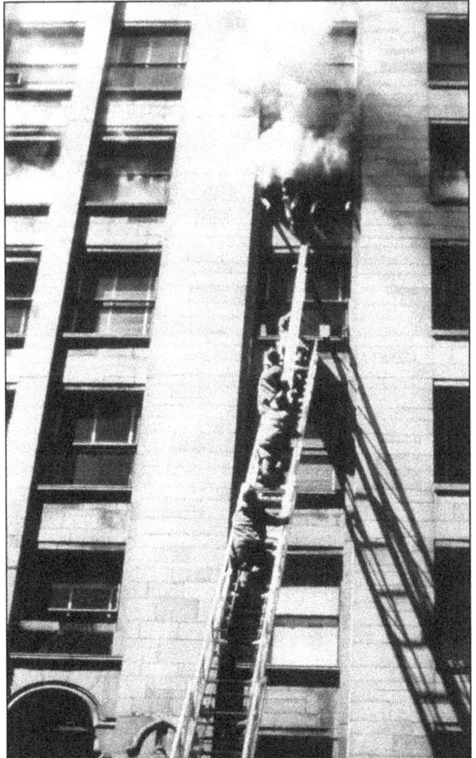

When Ladder Company No. 1 pulled up, firefighters could see people hanging out of windows on the upper floors of the Buhl Building. Some victims were on the street side, while others were trapped in rooms overlooking a roof, four floors up at the side of the building. Ladder Company No. 1 was hampered by double-parked cars. When they extended the 100-foot aerial ladder, they were one floor short of the trapped victims. The captain ordered a ladder-splice operation. (Courtesy of Matt Lee.)

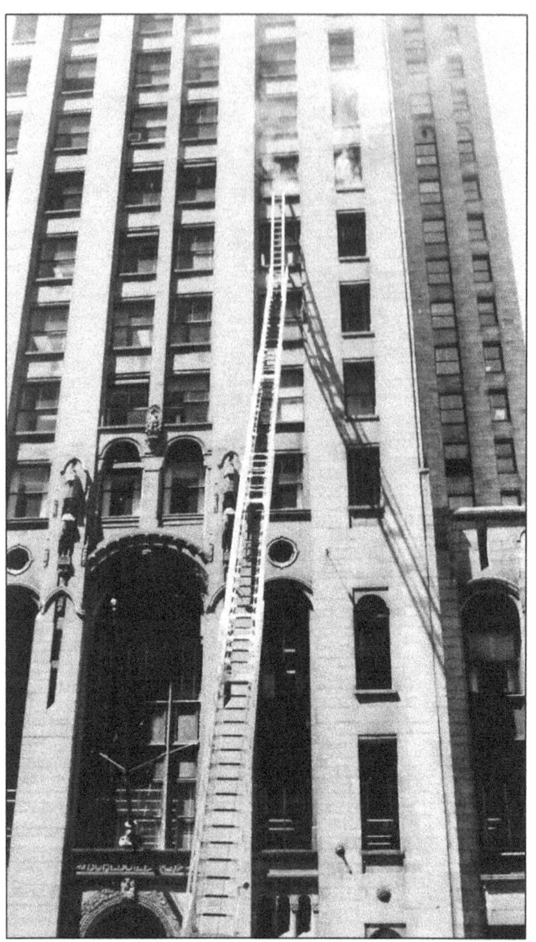

This is a rare tactic, but it was necessary under the conditions at the scene. Several firefighters took up a 16-foot roof ladder and placed it on the upper rungs of the aerial ladder. The roof ladder was secured with leather belts. The men proceeded to help victims out of the windows and down a precarious setup eight floors above the street. (Courtesy of Matt Lee.)

On January 15, 1979, a fire broke out at Six Mile Road and Carrie Street in a commercial building, which escalated to a second alarm. During operations, a firefighter went down in the building. Here, members from Engine Company No. 56 help drag out the injured fireman to safety and medical attention. (Courtesy of Bill Grimshaw.)

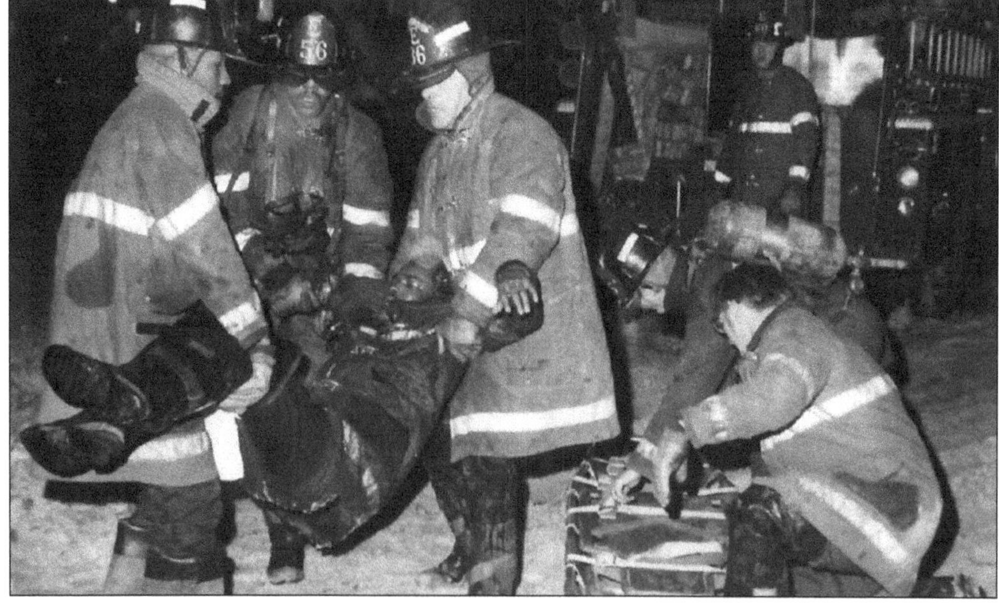

Detroit firefighters are trained to perform search and rescue. Here, at a dwelling fire located at Helen and Miller Streets, members from Squad Company No. 6 came upon this civilian who was suffering from smoke inhalation. A firefighter administered oxygen to the young man in an effort to render first-aid until Detroit EMS arrived. (Courtesy of Bill Grimshaw.)

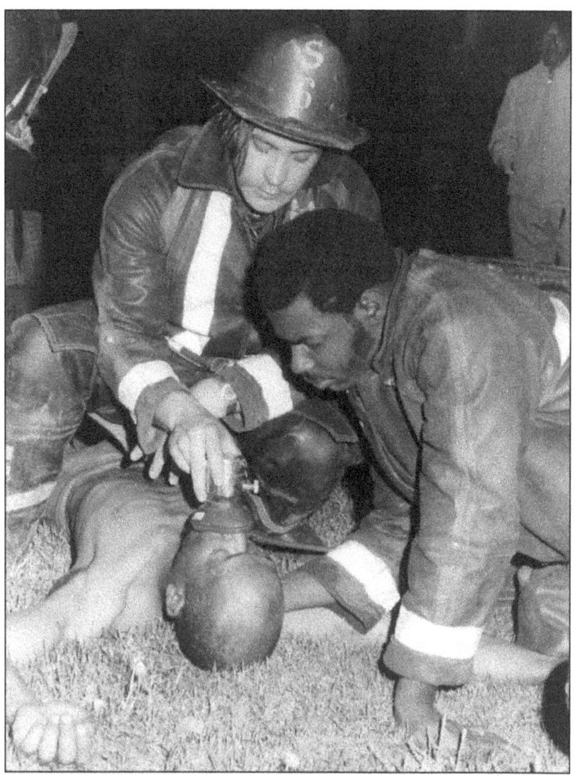

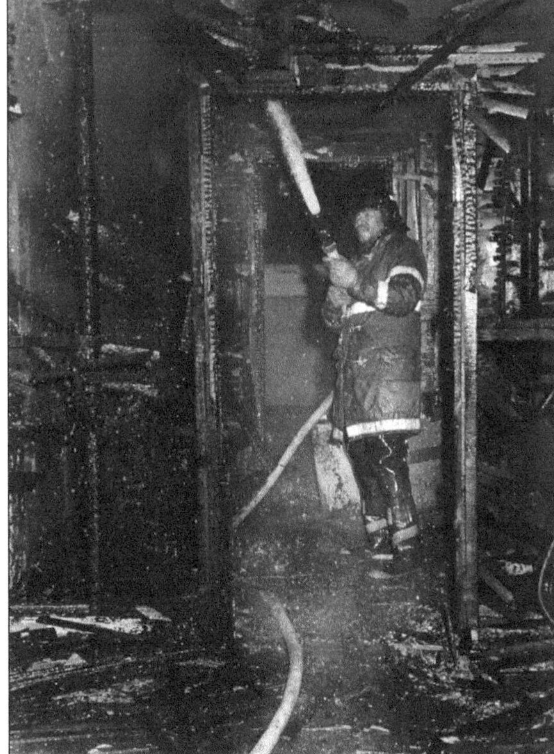

A firefighter conducts a final wash down of the interior of this structure fire at Proctor Street and McGraw Avenue. Using a one-and-a-half-inch hose line with a fog-pattern nozzle, a final wash down applies water to hot, burned wood so no fire can reignite and burn again. (Courtesy of Bill Grimshaw.)

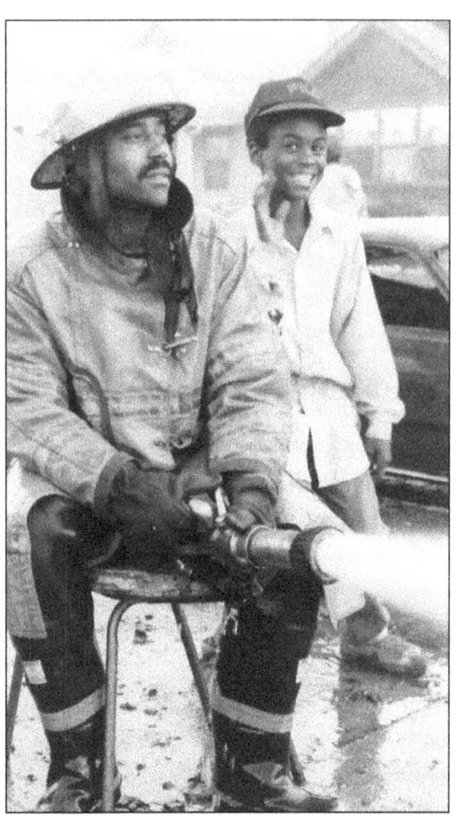

A Detroit firefighter is at a multiple-alarm fire at Monica Street and Westfield Avenue; he is using a hose line to extinguish flames from the building. He managed to find a chair that allowed him to work and catch a break at the same time. Note the young boy examining what the firefighter is doing and posing for the camera. (Courtesy of Bill Grimshaw.)

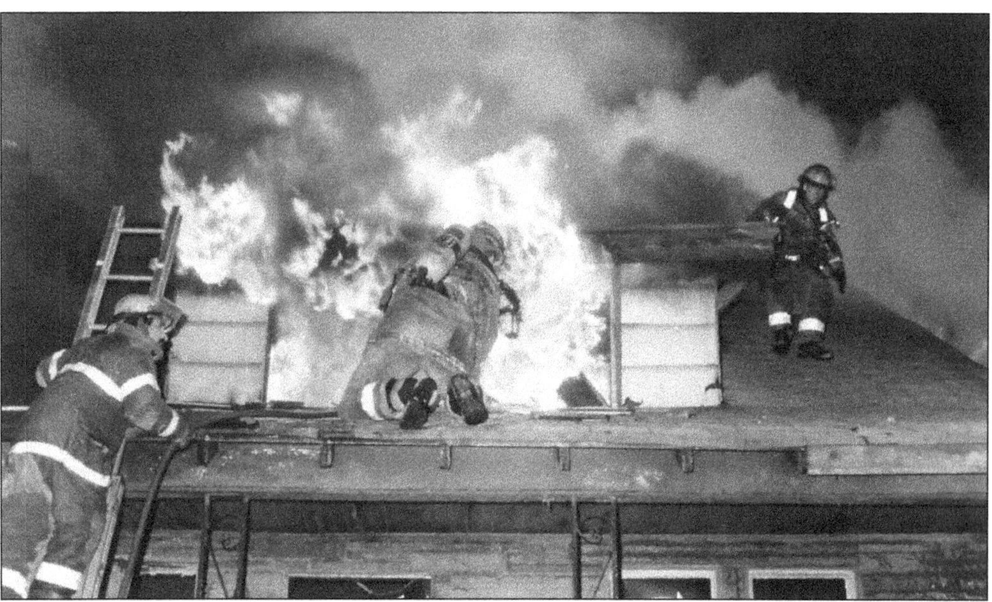

Devil's Night fires can always become spectacular events. Here, a fire in a vacant, two-story dwelling with no stairway inside the house forced firefighters to attack from the exterior with the help of ladders. In doing so, the firefighter on the roof had break the window to use it as an entrance point. This introduced air to the blaze, which caused it to flash, and the firefighter became engulfed in flames. (Courtesy of Bill Grimshaw.)

Seven

Detroit Today

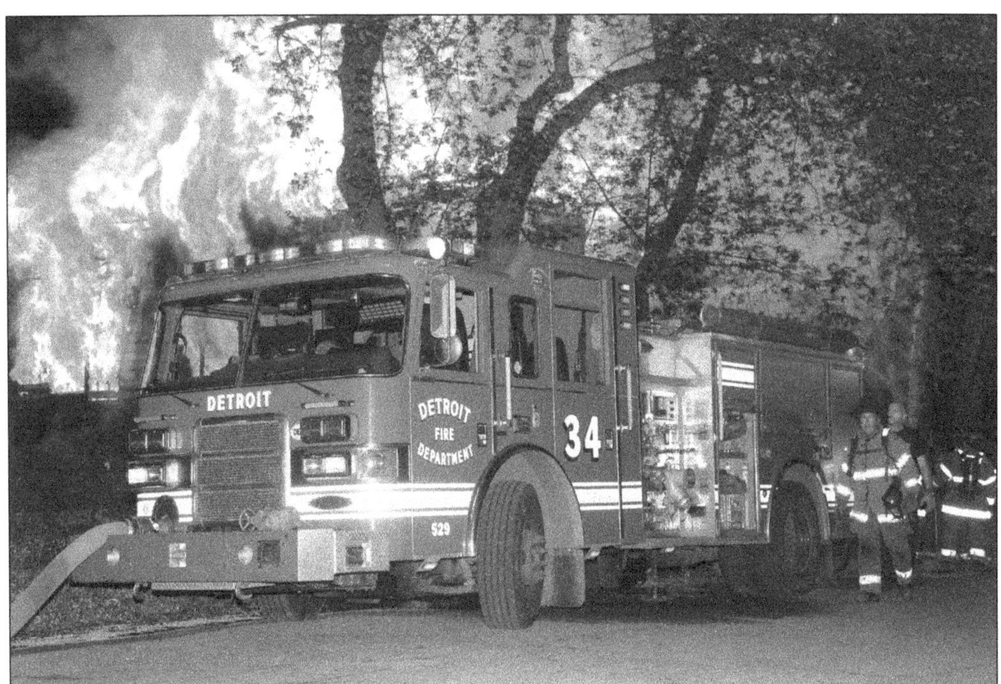

Detroit has seen major ups and downs over the years. The population of the city has seen a sharp decline in recent decades, capped by the City of Detroit declaring bankruptcy in 2013. Fires caused by arson have risen to where fire buffs from across the United States and Canada travel to Detroit on a regular basis to set up cameras and video equipment to record the destruction and observe fire operations. (Courtesy of David Traiforos.)

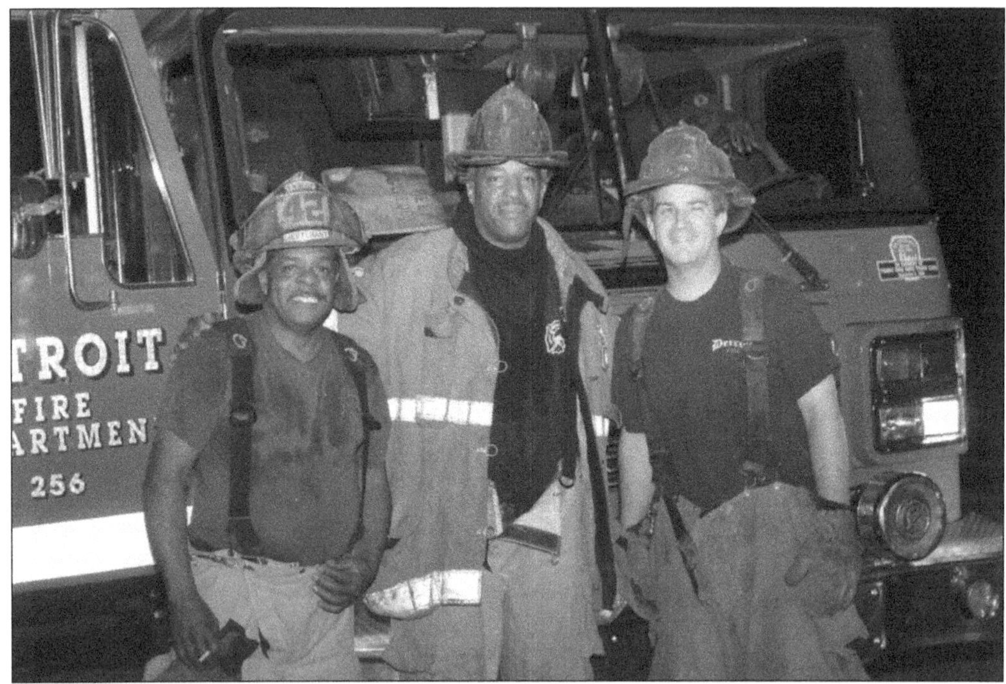

Detroit firefighters spend 24 hours a day working with each other every other day. They become much like a family. They rely on one another and help each other in serious fire conditions and in matters of life and death. (Courtesy of David Traiforos.)

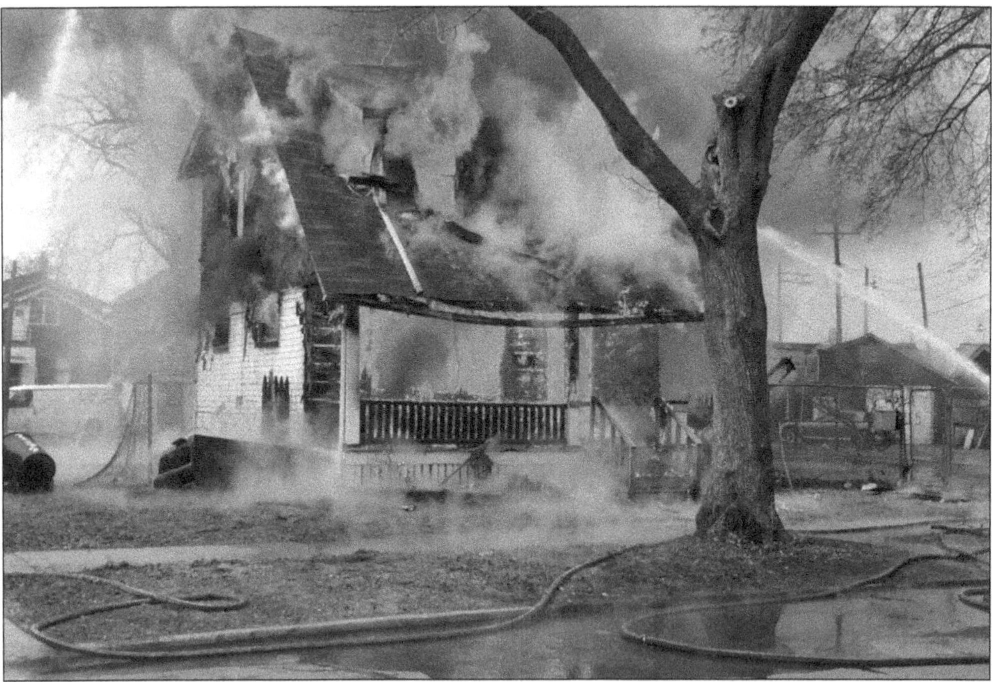

It does not have to be Devil's Night to see vacant dwellings burning during daylight hours. Fires like this are a common site throughout the city. The Detroit Fire Department is busy responding to several fires like this on a daily basis. (Courtesy of David Traiforos.)

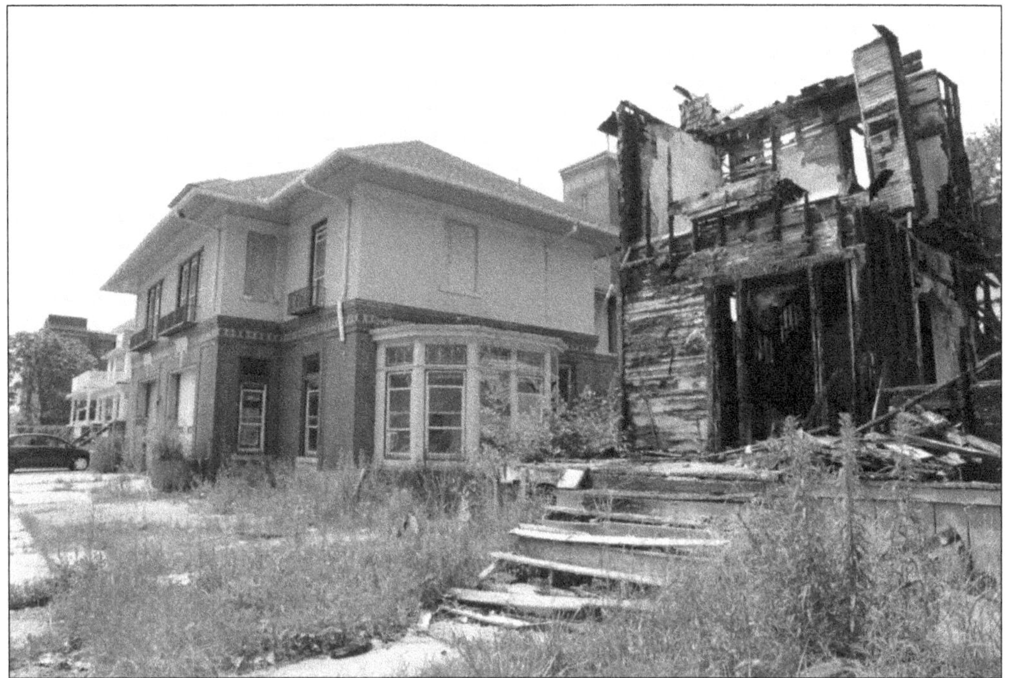

This is a view of the abandon Engine Company No. 38 firehouse on Detroit's east side; it sits next door to a burned-out vacant dwelling. Both vacant buildings were vandalized, and any building materials of value were stolen. (Courtesy of David Traiforos.)

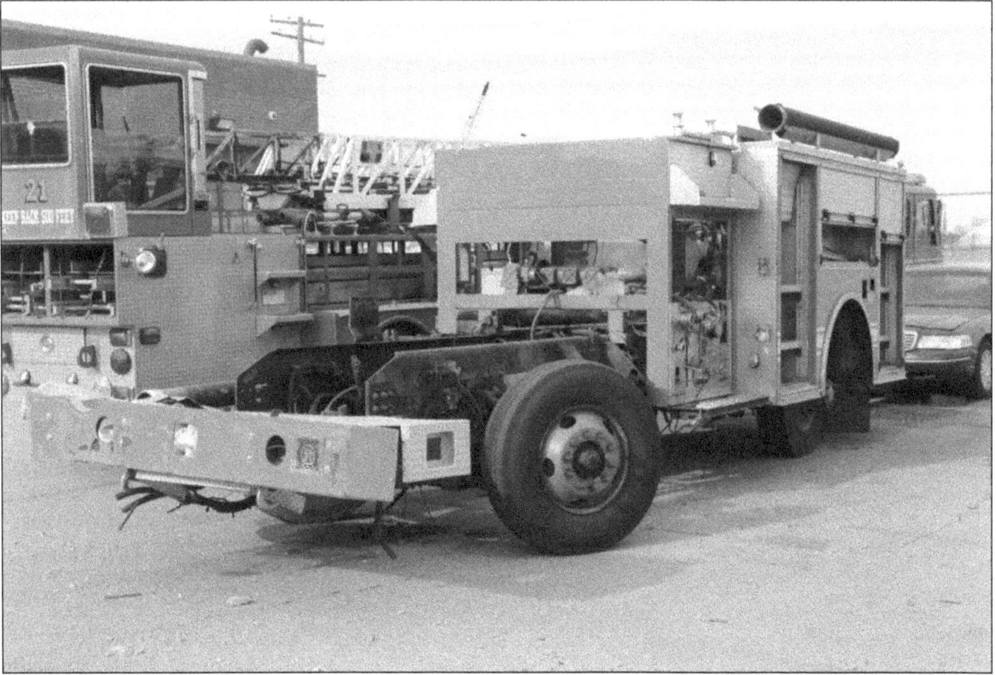

The financial issues of the City of Detroit have trickled down to the fire department. Here, a vehicle no longer in service at the repair shops donated its workable parts to other vehicles in need of repair. (Courtesy of David Traiforos.)

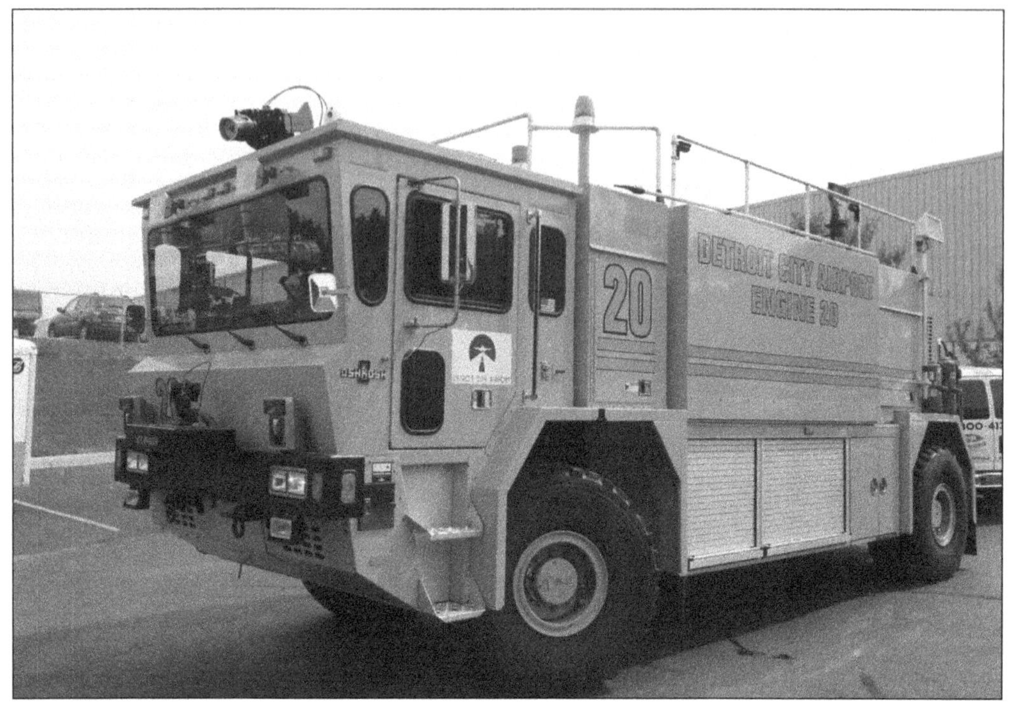

Here is a view of Engine Company No. 20, located at the Detroit City Airport. This Airport Crash Fire Rescue (ACFR) vehicle was taken out of service. Currently, the apparatus is being overhauled, and the city plans on reopen Engine Company No. 20 again. (Courtesy of David Traiforos.)

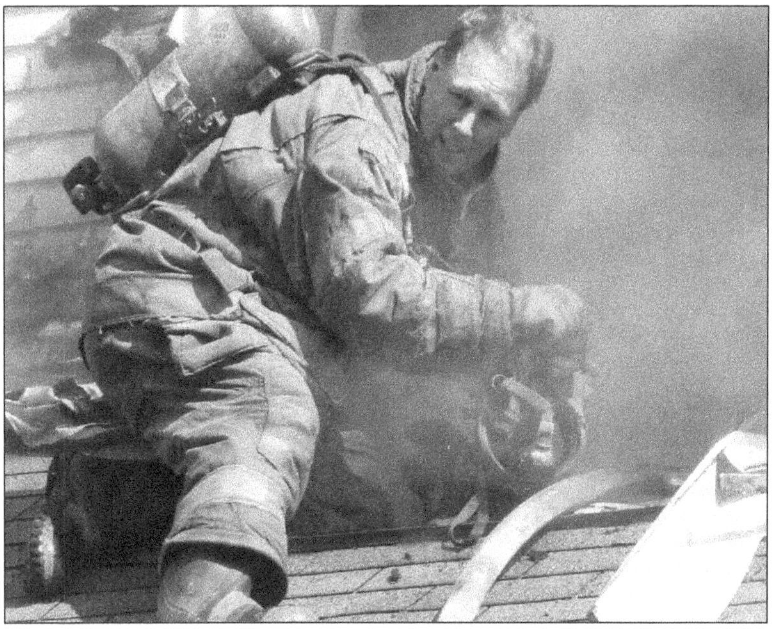

Lt. Pat O'Dowd is seen working at a fire; he is on the roof of the dwelling, preparing to mask up while waiting for water to charge the hose line before making entry into the structure through a window. (Courtesy of David Traiforos.)

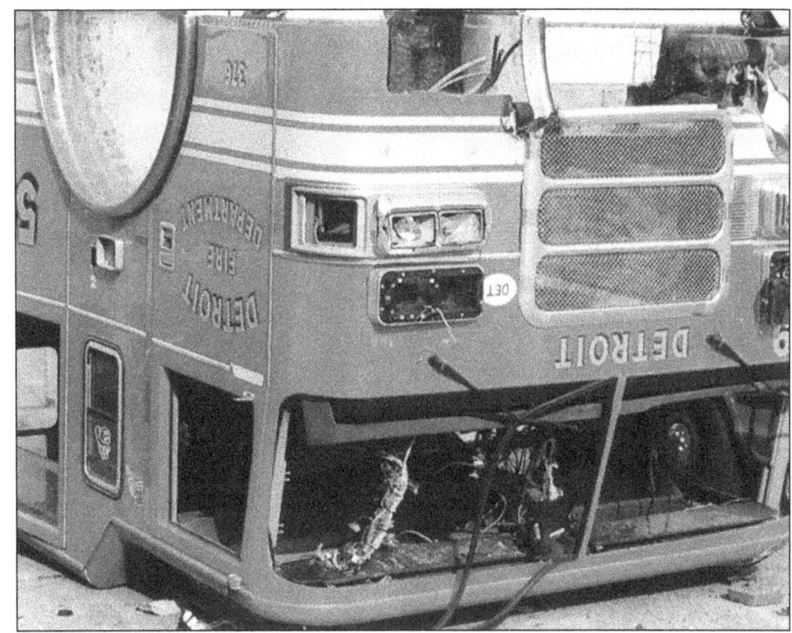

This is a scene from the repair shops on Russell Street. Here, the cab of Engine Company No. 59 was replaced, and the old cab was used for parts for other vehicles as needed. (Courtesy of David Traiforos.)

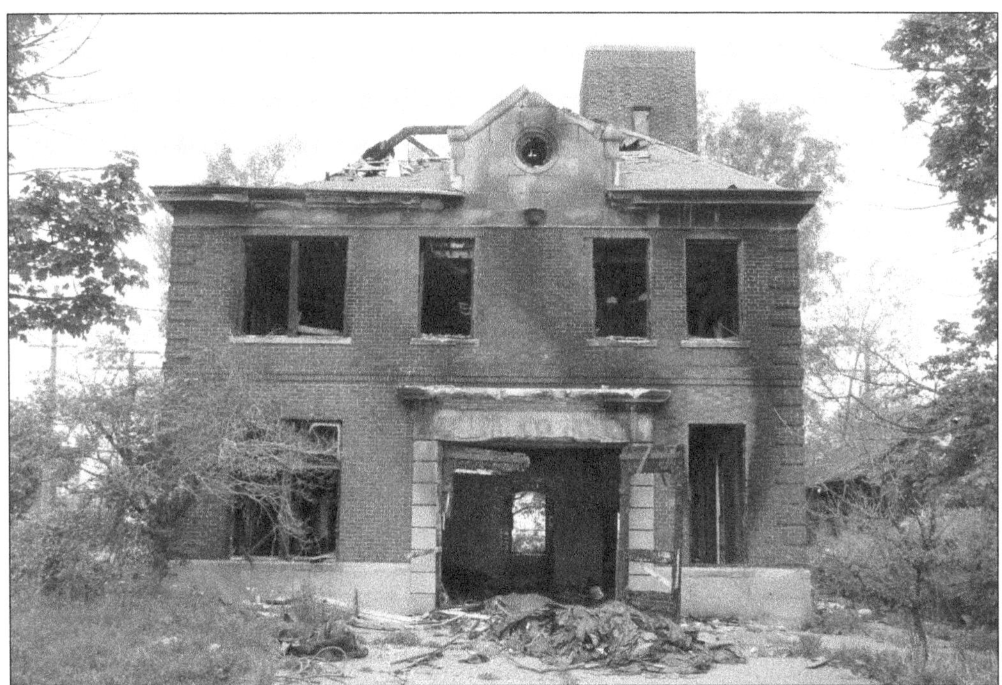

The former quarters of Engine Company No. 48 was located at Bayside and Saunders Streets in far southwest Detroit. This firehouse opened in 1924 and closed in 1980. It was privately owned and lived in until fire struck and destroyed the structure. (Courtesy of David Traiforos.)

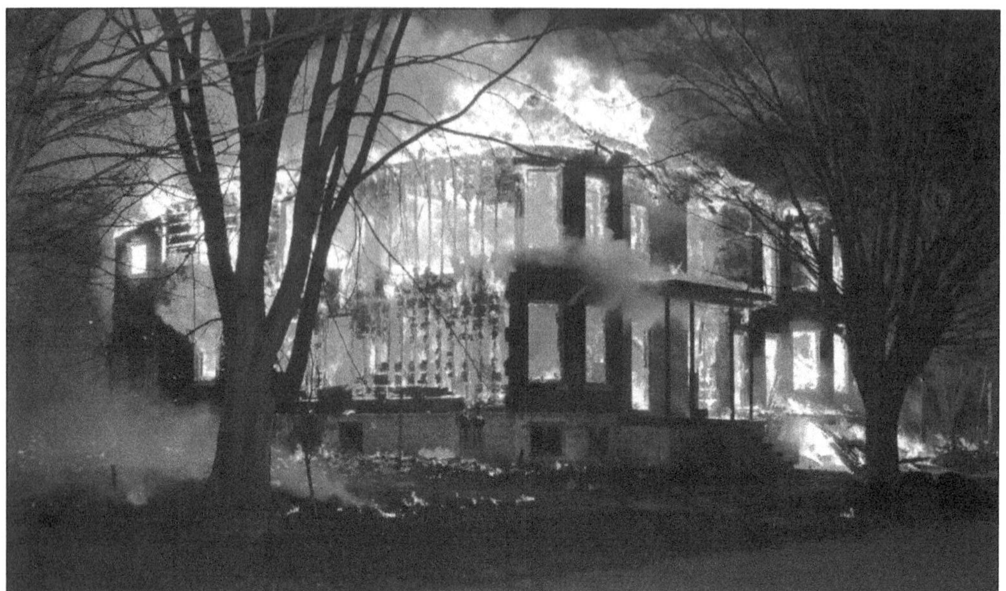

When residents leave and neighborhoods are abandon, the businesses and houses do not go with them. This creates a ripple effect and decimates a once thriving neighborhood. Detroit was built for two million people, and the population is now close to 600,000. The city has launched a demolition project to remove these vacant buildings. This fire scene shows multiple vacant homes burning during the middle of the night. (Courtesy of David Traiforos.)

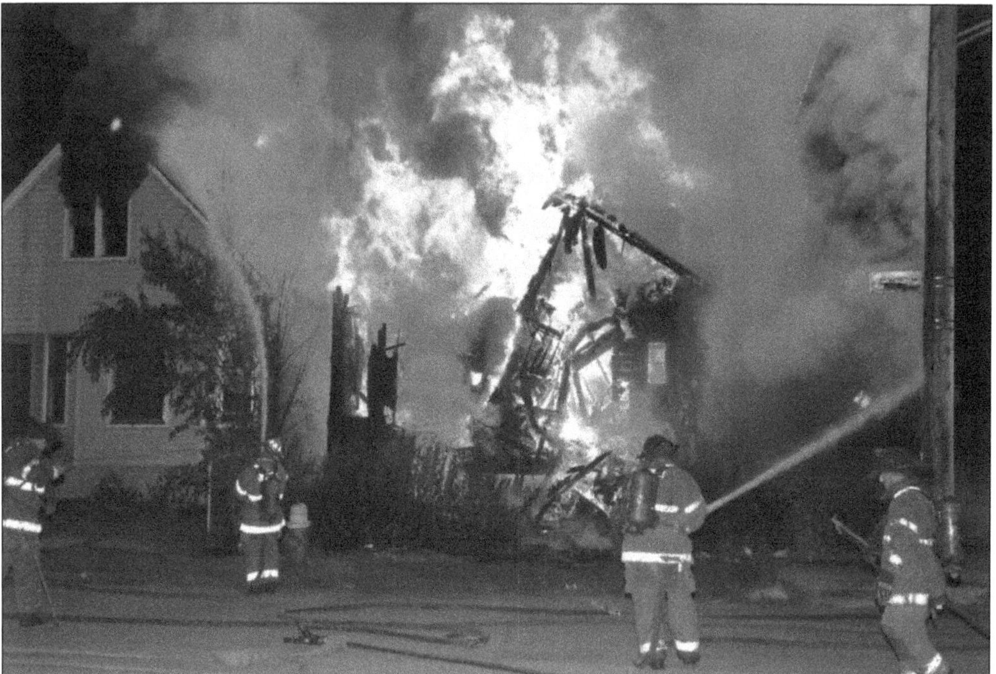

A fire lit in a vacant home by an arsonist can quickly spread to other homes either vacant or occupied. Firefighters respond to each and every call for help. The fire department may have cuts in many areas, but not in serving the public and responding to calls for help. (Courtesy of David Traiforos.)

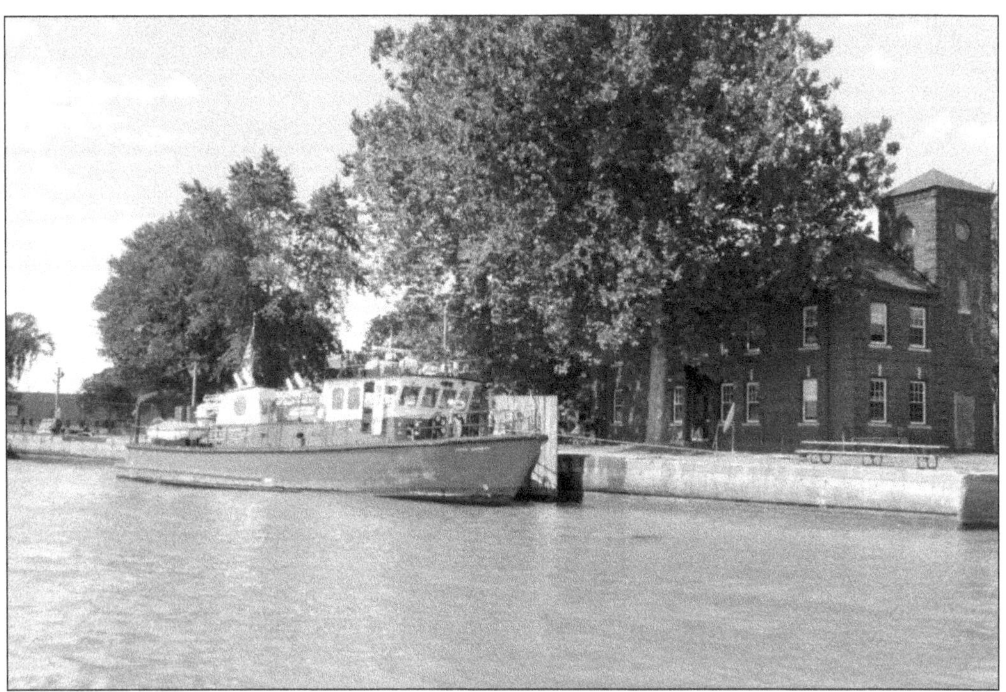

The Detroit fireboat is also referred to as Engine Company No. 16. This boat is docked on the Detroit River at the foot of West Grand Boulevard. When put in service, it was painted lime-yellow but later was repainted red. The boat is named after trail firefighter Curtis Randolph of Engine Company No. 32, who was killed in the line of duty on October 29, 1977, from injuries sustained in a back draft. This view shows the boat and its quarters. (Courtesy of David Traiforos.)

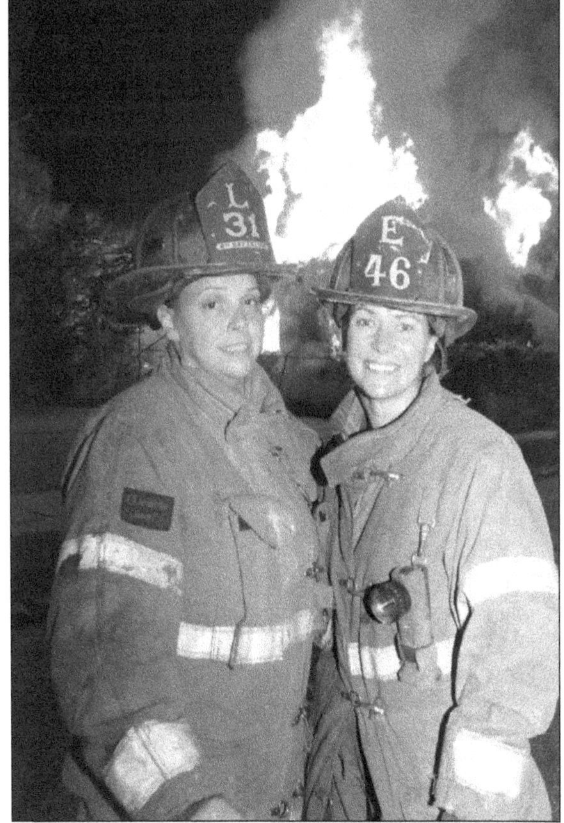

The word "brothers" is commonly heard in the Detroit Fire Department, describing the relationship firefighters have with one another—but "sisters" also work together at the Detroit Fire Department. (Courtesy of David Traiforos.)

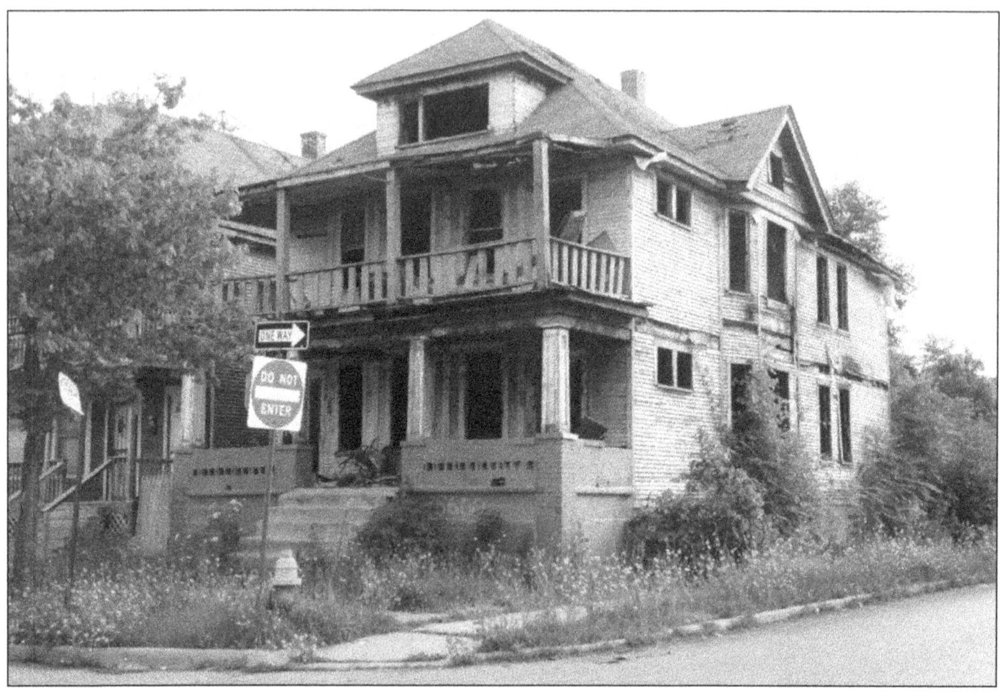

The corner of Campbell Street and McGraw Avenue is seen around noon on July 13, 2008. Here sits what once was a beautiful two-family residence. Tough times cause residents to abandon their homes and move on. (Courtesy of David Traiforos.)

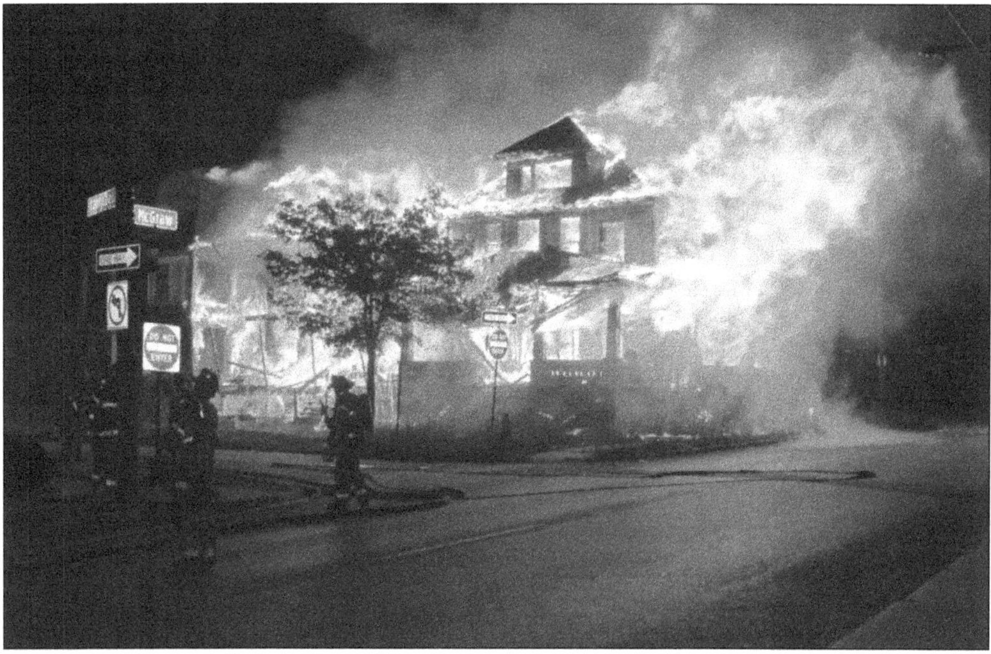

The corner of Campbell Street and McGraw Avenue is seen around midnight. Here sits what once was a beautiful two-family residence. The fire quickly spread to two other vacant homes nearby. The heat from this fire caused severe damage to four other homes across the street. (Courtesy of David Traiforos.)

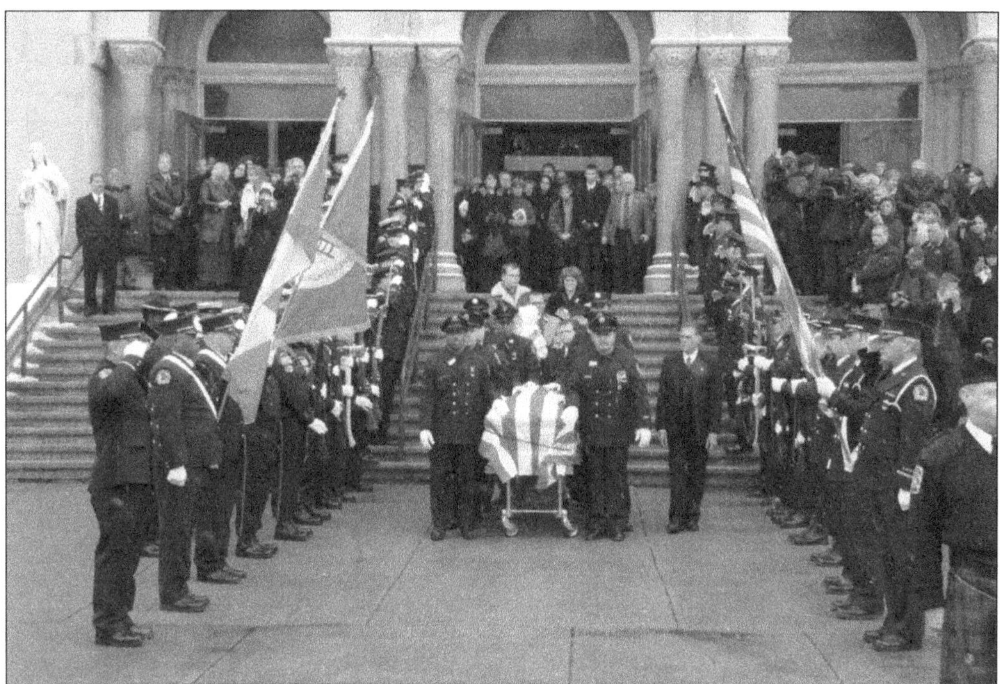

On February 7, 2007, FEO Joseph Torkos of Engine Company No. 17 was killed in the line of duty from injuries sustained in a vehicle accident. This view shows the pallbearers carrying his flag-draped casket out of St. Hedwig Catholic Church between two Honor Guards, one from Detroit Fire Department and the other from the Windsor, Ontario, Canada Fire Department, onto a fire department engine with his family walking behind. While snow fell, members of the Detroit Fire Department Band played New Orleans–style jazz music that could be heard throughout the neighborhood, which consisted of many burned buildings. The schoolchildren from St. Hedwig School came out to pay their respects. The funeral was attended by hundreds of firefighters from all across the United States and Canada. (Courtesy of David Traiforos.)

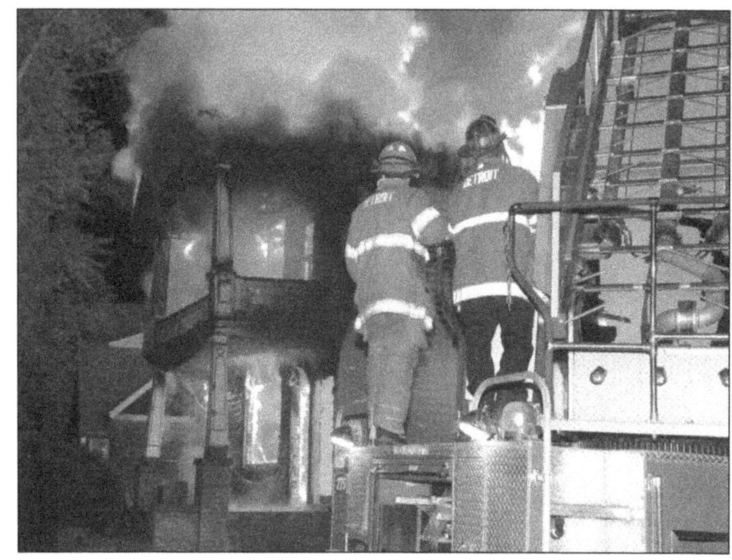

Firefighters from Ladder Company No. 13 pour water from their raised and extended aerial pipe onto a fire in a vacant two-family dwelling at Prairie Street and Diversey Avenue. A large volume of fire quickly caused the wooden building to become weakened and collapse. (Courtesy of David Traiforos.)

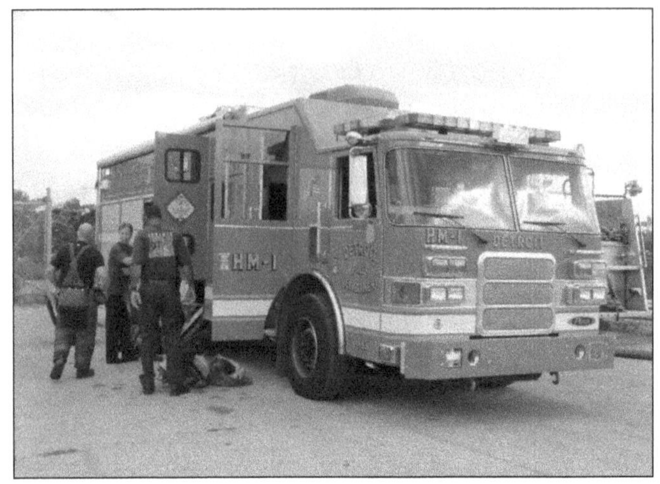

Detroit, once a leader in the fire service, still manages to keep up with the rest of the country in regards to technology while facing budget issues. The DFD established a hazardous materials unit. Here, the unit arrives at the scene of a suspected hazardous materials incident for a chemical spill. Members prepare to suit up and monitor air quality to establish a safe zone. (Courtesy of David Traiforos.)

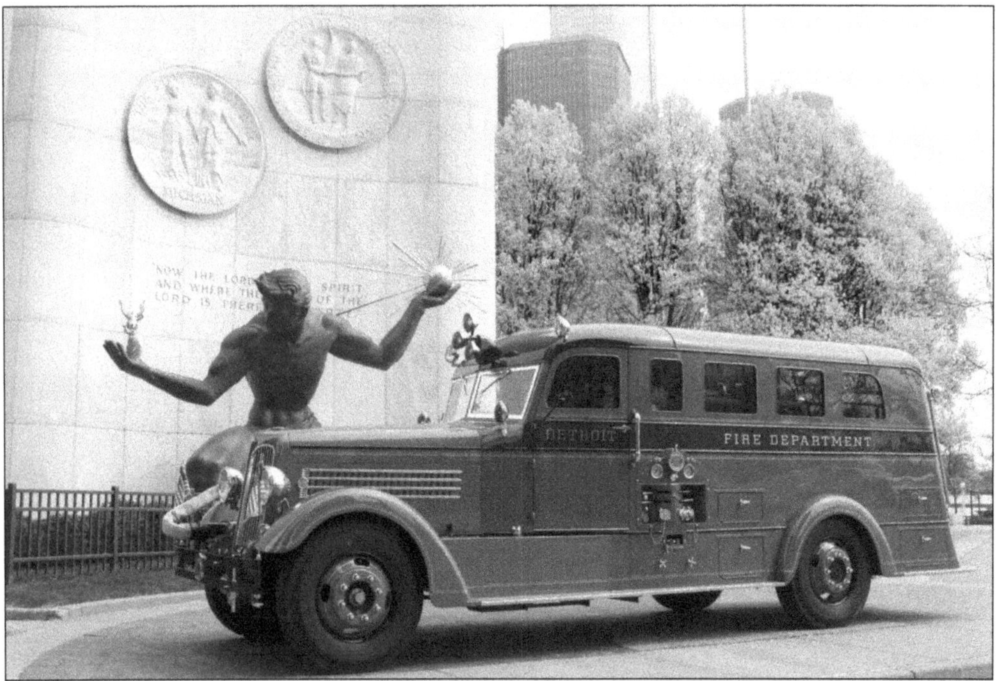

The Detroit Firemen's Fund took on an 11-year project to provide a final ride on a fire truck for Michigan firefighters who died in the line of duty; it is a tradition in danger of disappearing because modern-day fire trucks do not lend themselves well for this purpose. A 1937 Seagrave Safety sedan was refurbished from top to bottom, from the front bumper to the rear back step. The funds for this project were gathered from fundraisers in addition to collecting monetary, time, and material donations from large corporations and small businesses. Here, the restored memorial rig stands next to the Coleman Young Municipal Building in downtown Detroit. (Courtesy of David Traiforos.)

A firefighter using a hose line darkens down flames in a vacant commercial building fire to venture farther into the structure while his officer from Engine Company No. 42 watches for both members' safety while observing the damaged building's condition. (Courtesy of Bill Eisner.)

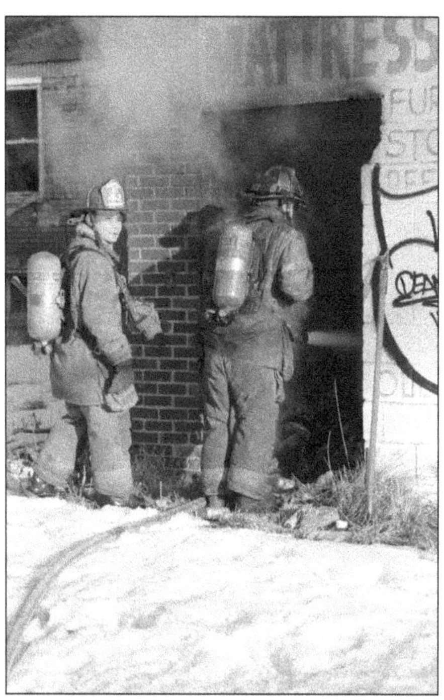

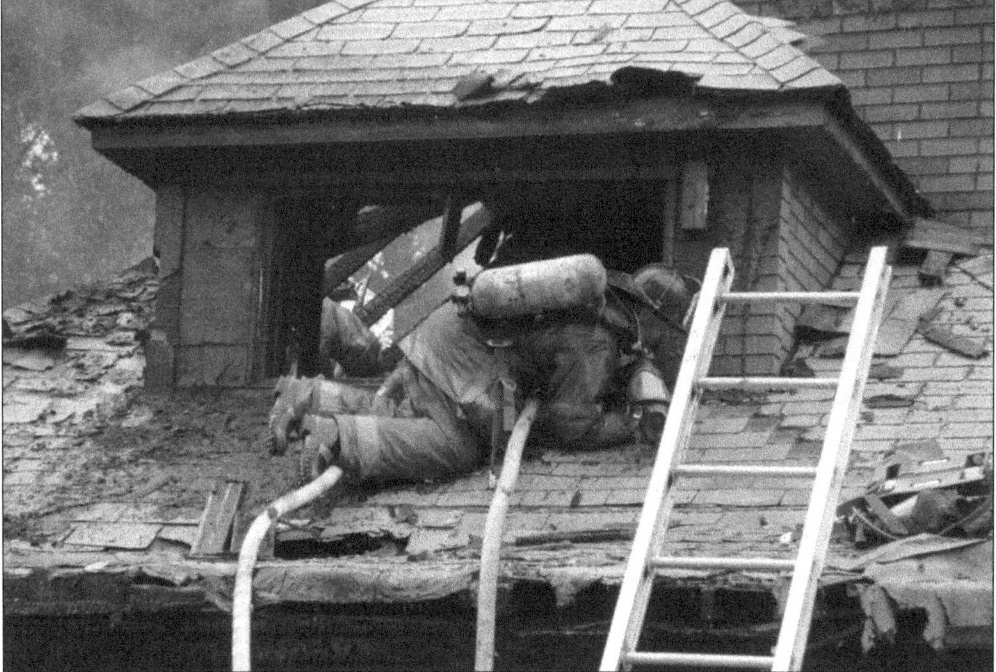

Fires in vacant buildings can lead to injuries to firefighters. At this vacant dwelling fire, the vandals removed the interior stairway. Fire on the second floor had to be fought by firefighters raising portable ladders to the second-floor front window to gain access and advance their hose lines. The fire conditions on arrival had extended to and severely damaged the roof that the firefighters were standing on. Here, a firefighter begins to fall into the weakened roof while trying to pull himself away from the edge. (Courtesy of David Traiforos.)

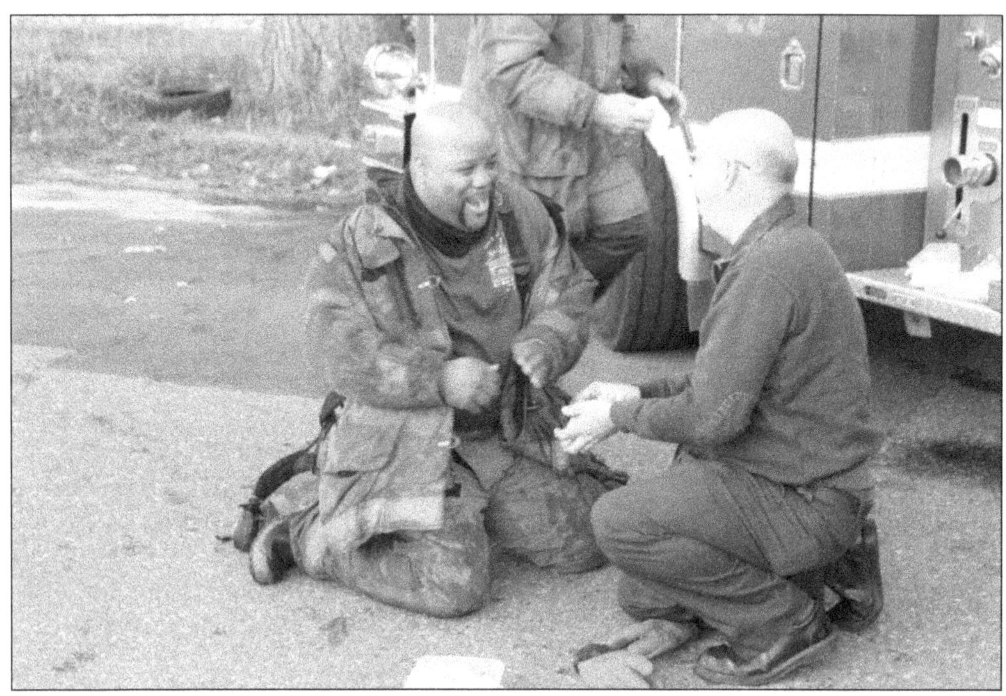

One of the most beloved firefighters of the Detroit Fire Department for generations, Senior firefighter Walter Harris of Engine Company No. 23 on Detroit's east side is being tended to at a fire by FEO Dave Springer for a cut on his head. This was one of the last photographs taken of Walter. He always had such a wonderful smile. He died shortly thereafter, killed in the line of duty from injuries sustained in the collapse of a roof after a fire in a vacant dwelling. Detroit police investigated the case, and the fire was ruled an arson-for-profit scheme. Walter left behind a wonderful family. His son James is a Detroit firefighter, following in his father's footsteps. In his spare time, Walter was a preacher who traveled from church to church on his motorcycle with his gun at his side due to the areas he served. He is deeply missed by all those whose lives he touched. (Courtesy of David Traiforos.)

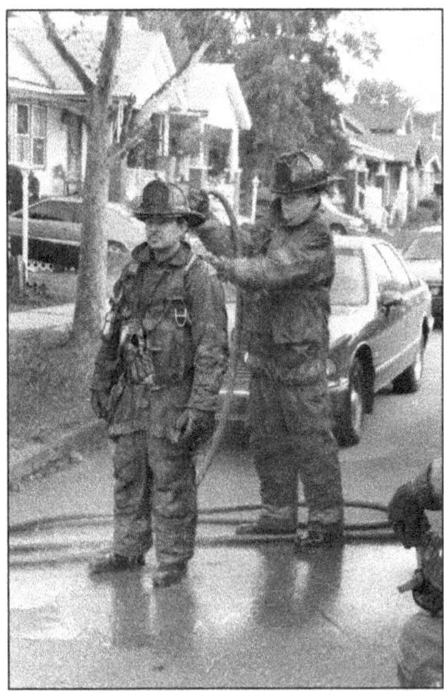

At this fire scene, a decontamination operation, washing off hazardous materials, takes place. A firefighter uses water to remove questionable materials off of another firefighter's gear. (Courtesy of Bill Eisner.)

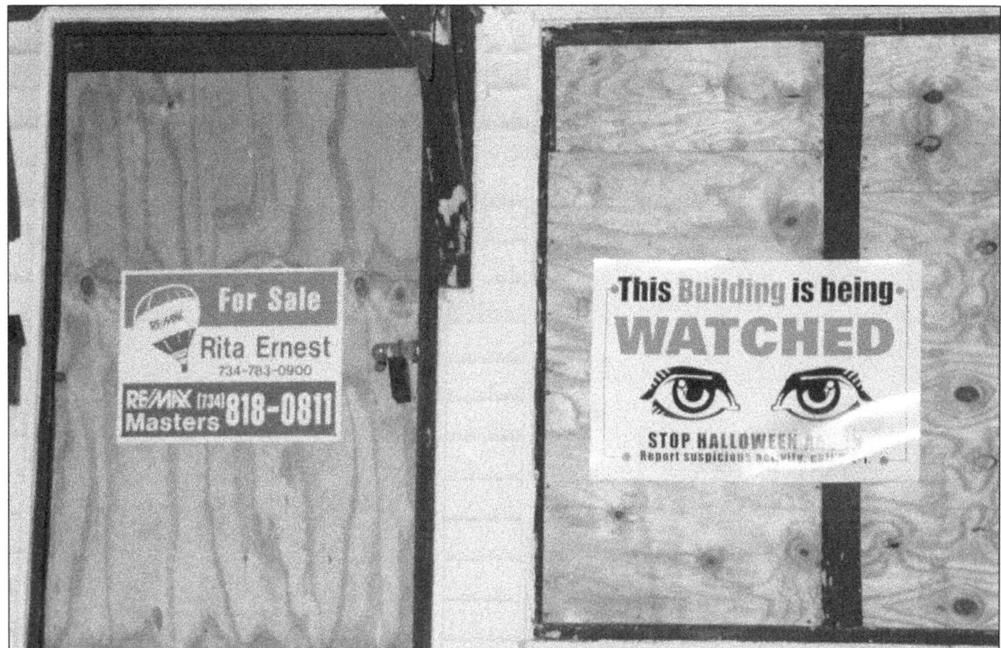

Prior to Devil's Night, Detroit firefighters take to the streets to distribute and post surveillance signs on vacant dwellings. These signs serve as a warning to would be "devils" that neighborhood "angels" are standing sentinel to keep watch over their neighborhood. (Courtesy of David Traiforos.)

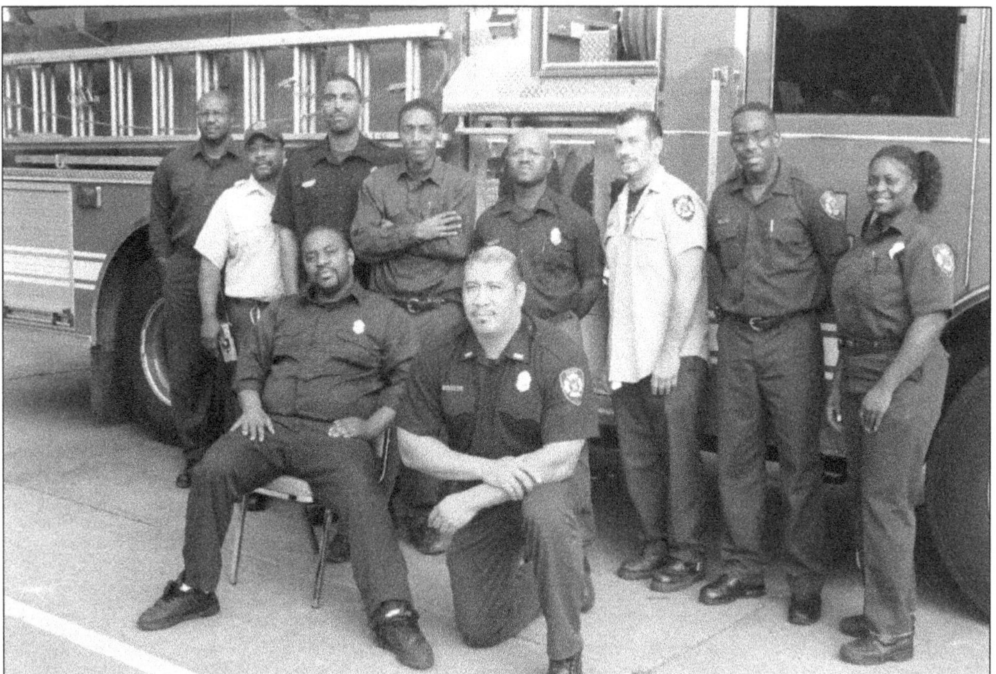

"Family" is what firefighters call each other. Here, members of Engine Company No. 40, Ladder Company No. 17, and Squad Company No. 5 pose for a family photograph for their scrapbooks. (Courtesy of David Traiforos.)

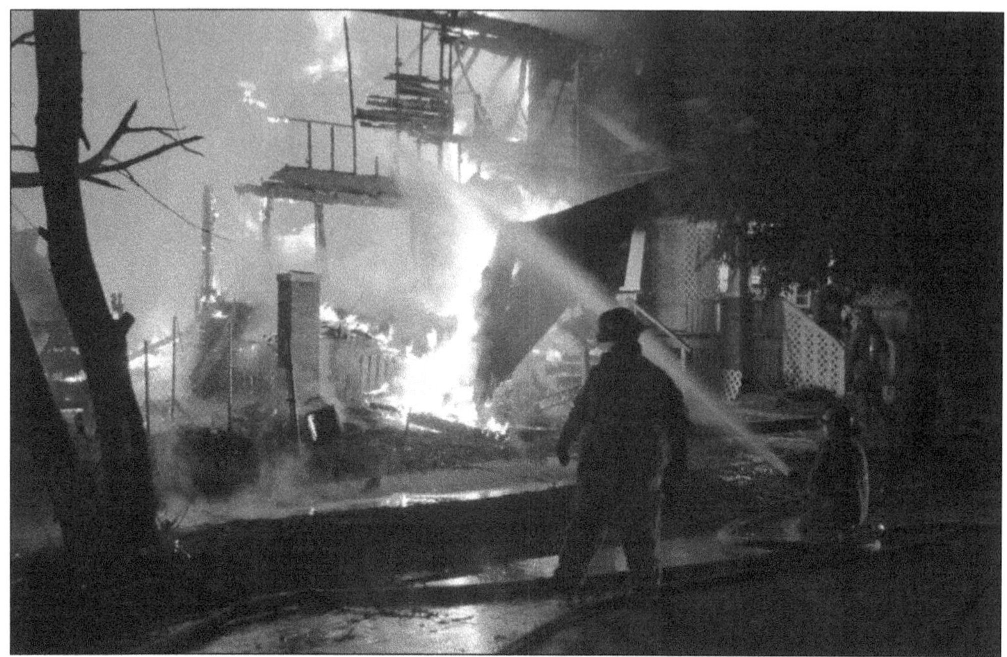

Fire to the left! Fire to the right! That is what Detroit firefighters sometimes face as they arrive on the scene where the blaze has spread to structures on both sides of the main fire building. Waiting for the driver of the fire engine to send water can sometimes seem like an eternity when trying to stop the spread of the fire. (Courtesy of David Traiforos.)

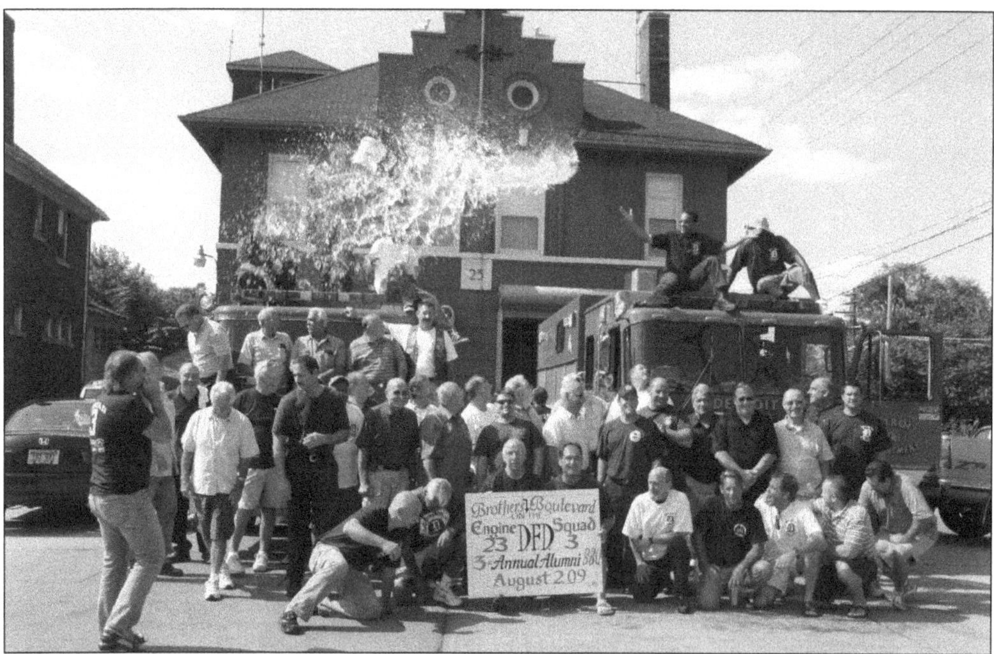

Every year, Engine Company No. 23 and Squad Company No. 3, located on East Grand Boulevard and Moran Street, hold their annual reunion of all the DFD members who used to run there. An annual photograph is taken, and it always seems to involve water. Engine Company No. 23 was closed in 2012 and taken out of service due to budgetary issues. (Courtesy of David Traiforos.)

The Detroit Fire Department has a corps of five chaplains who are there to serve the members of the DFD if spiritual matters are needed. Here, Allen McNeeley is giving a blessing to members of the DFD from Engine Company No. 31 for their 100th anniversary. Chaplin McNeeley was known for blessing members' motorcycles for safe travel, for wishing members well at retirement parties, partaking in Memorial Day events, and just talking to firefighters. McNeely passed away on October 26, 2013. (Courtesy of David Traiforos.)

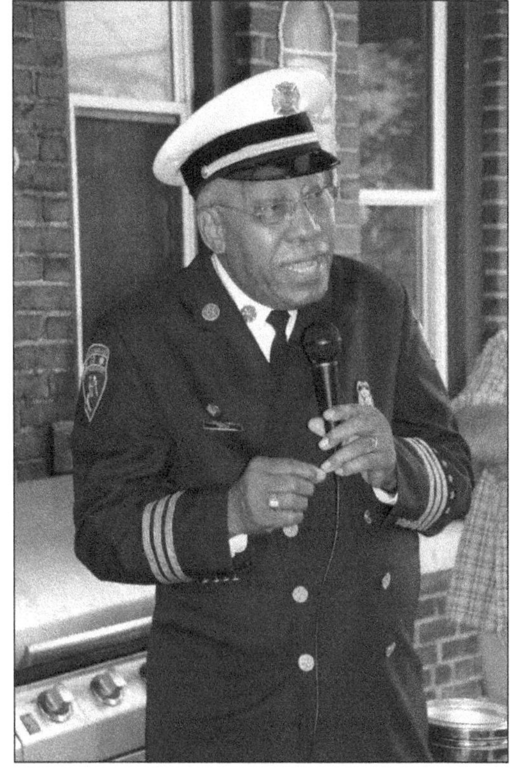

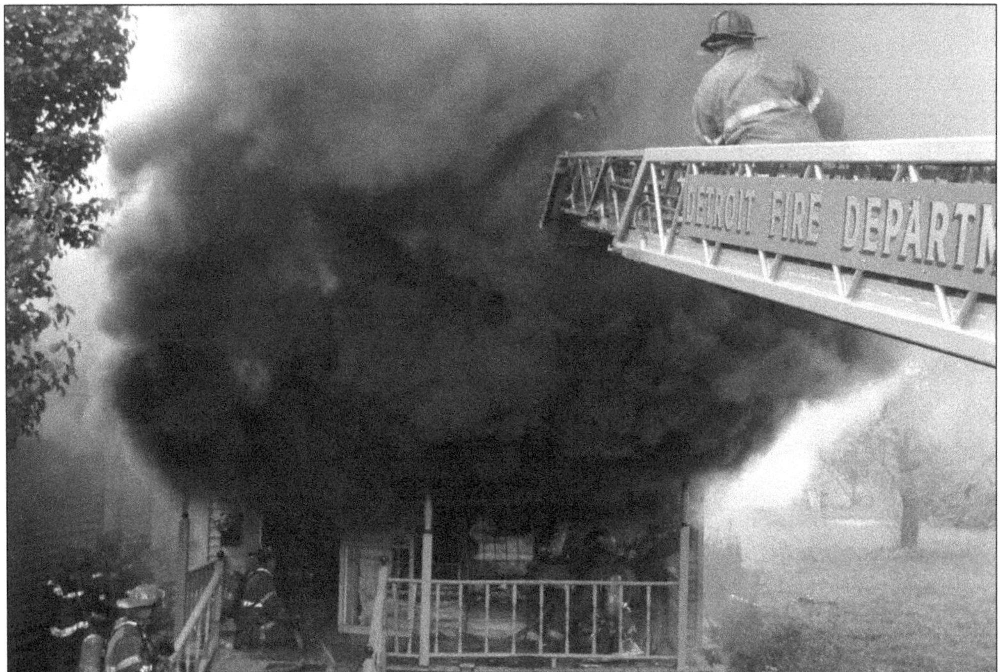

Here, at Seventeenth Street and Martin Luther King Drive, Detroit firefighters have their hands full with a fire in a dwelling. The smoky conditions have made it hard for firefighters to find the seat of the fire inside. This incident was one for the books. (Courtesy of David Traiforos.)

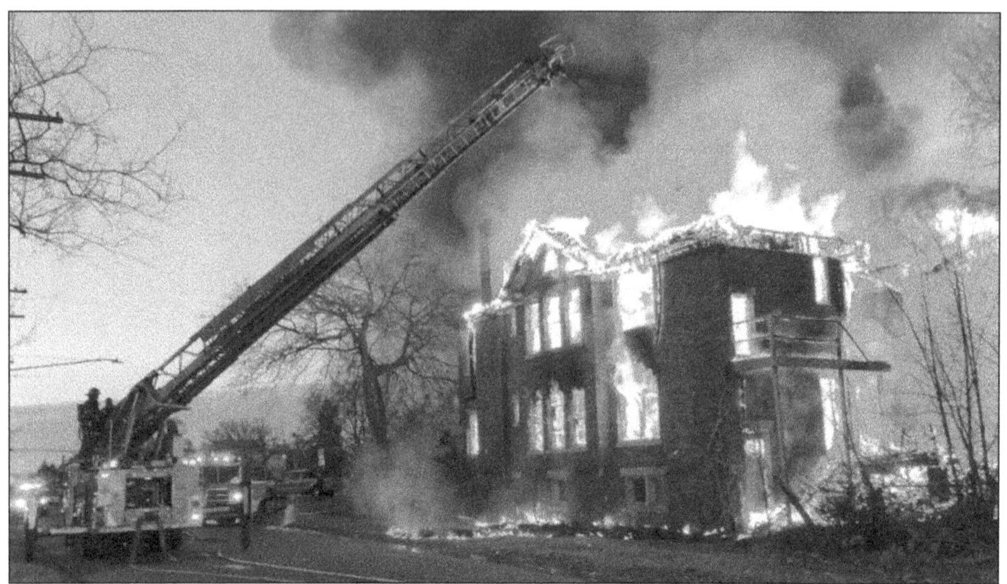

On March 22, 2015, Detroit firefighters responded to a report of a fire in a vacant apartment building. On arrival, they were faced with this fire; it quickly spread to another apartment building next door. Ladder Company No. 22 waits for water to flow through the deck pipe on the end of the aerial. (Courtesy of David Traiforos.)

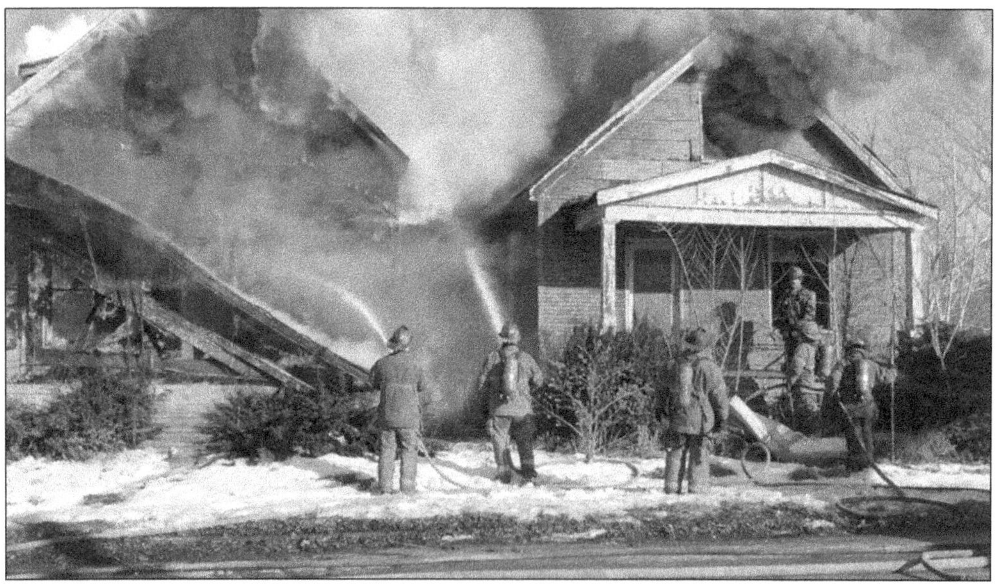

Firefighters arrive on the scene at Thirty-First and McGraw Streets to find fire in two vacant dwellings. They have taken a defensive stand on the building to the left, applying water streams from the outside. For the building on the right, firefighters make an offensive attack to enter the house with hose lines and attempt to extinguish the fire. (Courtesy of David Traiforos.)

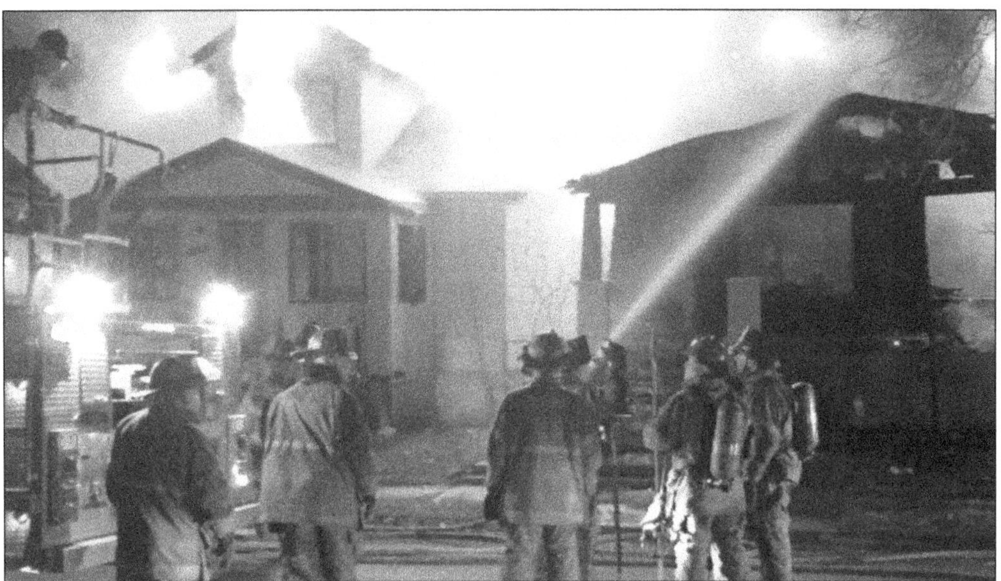

Firefighters responded to a report of a house on fire at Edsel and Omaha Streets. On arrival, they were faced with the challenge to extinguish fire in two dwellings and spreading to a third. Firefighters in many cases have limited manpower and equipment until they request additional units. The request takes time, for fire companies have to travel to the location. Sometimes, when it is busy with multiple fires happening at once, additional resources are not available, and firefighters have to make do. (Courtesy of David Traiforos.)

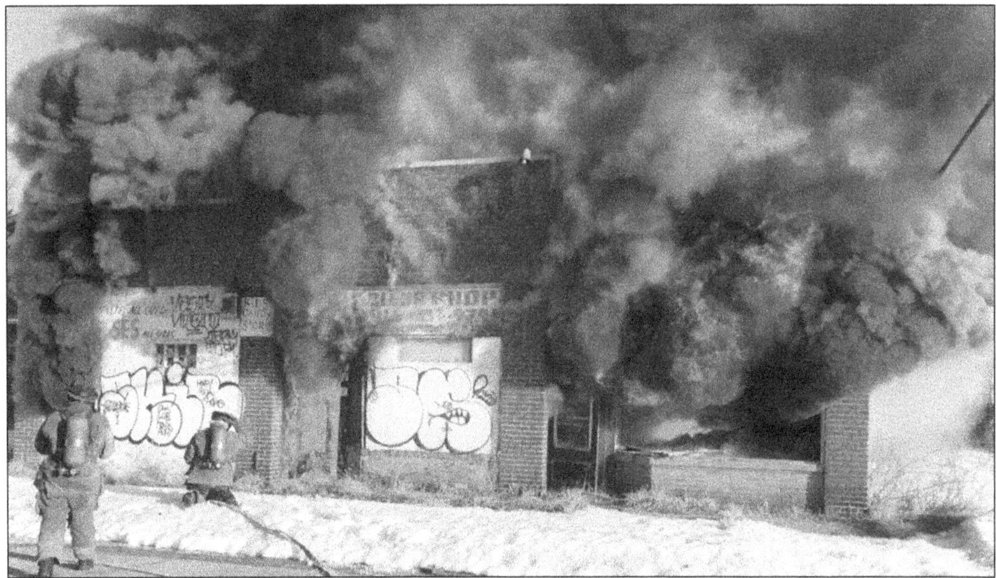

This fire at Twenty-Eighth and Milford Streets was in a vacant commercial building that spread to another vacant dwelling. Firefighters had some water issues in the initial stages of the fire due to the cold temperatures and a frozen hydrant. Shown are heavy fire conditions, which helps build up a firefighters' adrenalin, making this such an exciting job. (Courtesy of David Traiforos.)

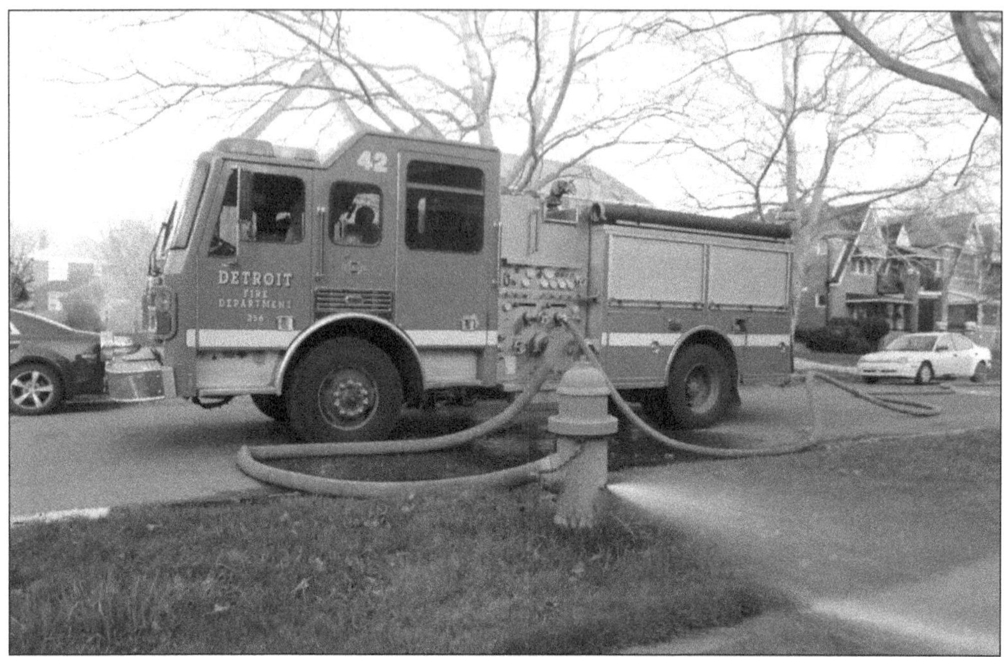

Detroit firefighters rely on the water system maintained by the City of Detroit. The hydrants supply precious water to help stop the spread of fire. The hydrant system in Detroit has helped prevent building fires from spreading into major conflagrations. (Courtesy of David Traiforos.)

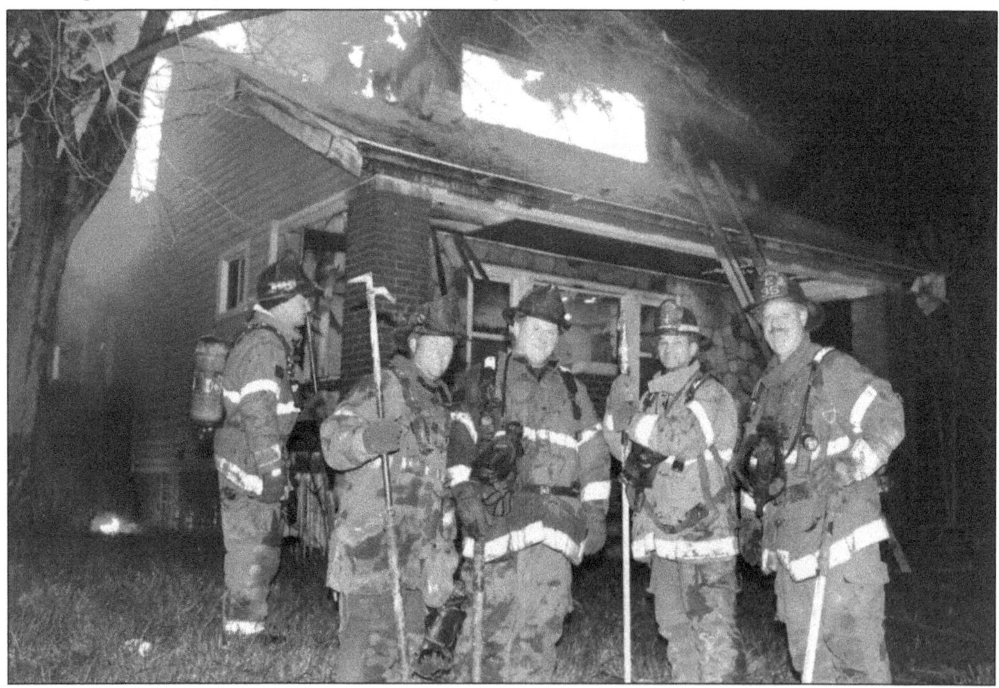
It is all about taking care of one another. Detroit firefighters work hard and play hard. They spend half of their life away from home with their fire family, working and exerting themselves for those they do not know, but when they call for help, the DFD responds. Being a Detroit Firefighter is a calling that all members can be proud of. (Courtesy of David Traiforos.)

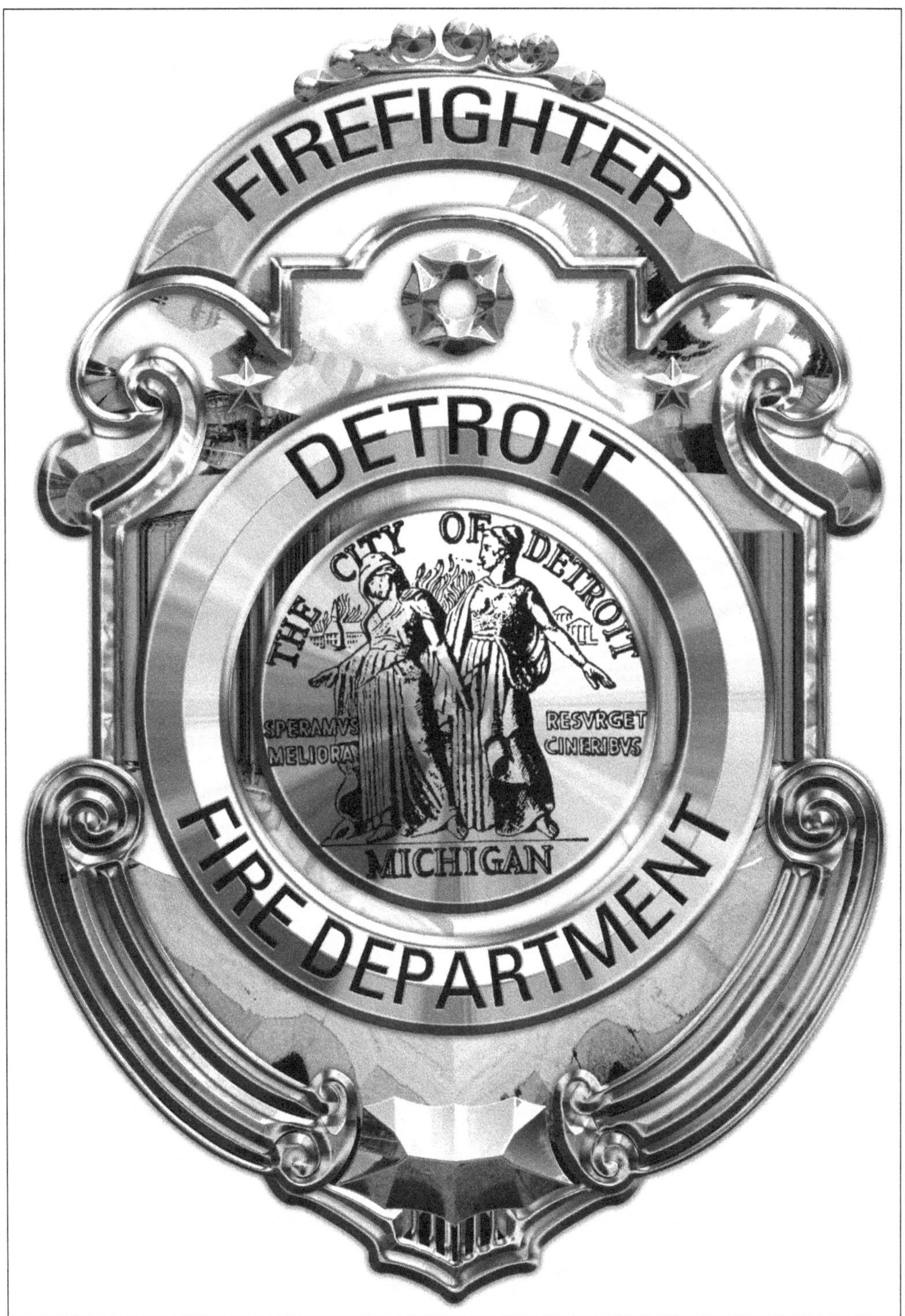

The badge is worn by all firefighters, officers, chiefs, and EMS members of the Detroit Fire Department. It is a symbol that is worn over their hearts. (Courtesy of Edward Abair.)

ROLL OF HONOR

The following Detroit Fire Department members sacrificed their lives in valiant service while protecting the lives and property of the citizens of the city:

08/17/1867—Pipeman John Miller E-3 Fell from Ladder
07/02/1881—Pipeman Michael McQueen E-8 Falling Wall
09/03/1884—Capt. William Cooper H & L 3 Injuries 02/26/1884
01/01/1886—Capt. Richard Filban H & L 3 Falling Wall
12/28/1888—Stoker James Powers E-8 Thrown from Rig
03/21/1889—Lt. Isaac Clark H & L 5 Floor Collapse
04/20/1889—Driver Henry Turner Chem 2 Thrown from Rig
12/03/1890—Pipeman Octavius Robinson E-8 Falling Walls
12/03/1890—Lt. Patrick Coughlin E-8 Falling Walls
02/28/1892—Pipeman Henry Chapoton E-5 Thrown from Rig
12/26/1892—Lad. Peter Schwartz H & L 2 Falling Walls
11/02/1893—Lad. Hugh Garrity H & L 3 Rig/Street Car Collision
11/11/1893—Eng. David Boyd E-4 Arm Torn Off
08/02/1894—Pipeman Eugene McCarthy E-19 Falling Walls
10/05/1894—Pipeman Martin Ball E-9 Falling Walls
10/05/1894—Pipeman Joseph Dely E-9 Falling Walls
10/05/1894—Pipeman John Pagel E-9 Falling Walls
10/05/1894—Lad. Julius Cummings H & L 2 Falling Walls
10/05/1894—Lt. Michael Donaghue Chem 1 Falling Walls
10/05/1894—Volunteer Fred Bussey Falling Walls
05/17/1895—Lad. Anthony Korte L-5 Rig Train Collision
01/13/1897—Lad. Patrick Black H & L 2 Electric Shock
01/20/1897—Lad. Moses Fortune H & L 1 Kicked by a Horse
02/25/1900—Lad. Timothy Keohane L-8 Falling Smokestack
10/25/1900—Lt. August Regentin E-11 Falling Walls
03/22/1902—Pipeman George Hough Chem 2 Injuries
12/08/1905—Lt. Michael Sheahan E-27 Pole Hole Fall
12/18/1905—Pipeman James Briggs E-3 Rig/Street Car Collision
03/09/1907—Cadet David Murduff E-3 Killed at Fire
11/16/1907—Driver James Downey L-4 Broken Skull/Run Over
02/11/1908—Pipeman Louis Staub E-14 Falling Wall
12/25/1908—Lt. William Burgess L-3 Died of Injuries from Fire
10/16/1909—Lad. John Wallace L-9 Died Driving Apparatus
01/14/1911—Capt. Levi Fletcher E-12 Falling Walls
08/08/1911—Pipeman Arthur Fitch E-24 Extinguisher Explosion
05/01/1913—Driver Timothy Shea E-34 Rig/Car Collision
01/31/1915—Pipeman Otto Habermas E-27 Fall from Ladder
07/19/1915—Sub. Willys Wilcox E-3A Run Over by E-3
12/27/1915—Pipeman William Hoffman E-22 Run Over by Engine
04/12/1916—Lad. William Moran L-3 Fell from Ladder
03/04/1917—Pipeman William Schill E-30 Building Collapse
03/04/1917—Pipeman Otto Mattick E-30 Building Collapse
03/04/1917—Lad. Oscar Locke L-1 Building Collapse
03/04/1917—Capt. Alexander Cockburn E-2 Building Collapse
03/04/1917—Pipeman Alonzo Raymond E-2 Building Collapse
10/15/1917—Sub. George Lloyd E-6 Fell through Pole Hole
05/10/1918—Capt. Oscar Reidel L-4 Fell through Floor
10/20/1918—Capt. Maurice Kelly E-32 Falling Walls
05/29/1919—Pipeman Peter Condry S-1 Gas Inhalation
05/30/1919—Pipeman William O'Brien S-1 Gas Inhalation
07/24/1920—Pipeman Louis Purol S-3 Rig/Street Car Collision
03/05/1921—Lad. James Thornton L-1 Injury/Hemorrhage
02/19/1922—Pipeman Louis Dique E-29 Burns
11/18/1923—Pipeman Herman Schulz E-46 Burns
04/18/1924—Lt. George Hawkins S-4 Rig Accident
04/22/1924—Pipeman Richard Beard S-4 Rig Accident
02/06/1926—Pipeman Stanley Doptis S-1 Rig/Street Car Accident
12/20/1928—TFF Vincent Rychik L-4 Rig Accident
03/29/1929—Lad. Edward Vernier L-8 Rig/Truck Accident
07/31/1930—FF Lief Christiansen E-24 Thrown from Ladder
01/22/1931—Lad. George Aylsworth L-8 Rig Accident
10/27/1933—Pipeman Frederick Stolp S-4 Rig/Car Collision
07/19/1935—FF Walter Sweeney S-3 Electrocution
08/21/1936—FF Joseph Hallman HP1 E-1/HP1 Collision
03/18/1937—Capt. Louis Pape S-1 Heart Attack/S-1 after Fire
10/24/1937—Lt. Russell Kelly L-7 Heart Attack at Fire
02/13/1939—Capt. Edward Mitten E-50 Smoke Inhalation
02/25/1940—FF Joseph Schneider L-22 Building Collapse
07/12/1942—FF Stanley Hausch E-17 Buried by Explosion
10/21/1944—Lt. Joseph Donnelly E-1 Heart Attack at Fire
01/05/1945—Sgt. Harry Shlotzhauer E-10 Heart Attack at Fire
03/12/1946—Lt. Oliver Strong E-52 Smoke Inhalation
01/16/1947—FF Charles Parish L-9 E-12 Explosion
01/16/1947—FF Paul Reiner L-9 E-12 Explosion
03/17/1947—FF Raymond Benedict S-3 Smoke Inhalation
04/04/1947—Sgt. Fred Bergman E-32 Heart Attack/E32 after Fire
06/29/1947—Lt. Clay Carpenter E-22 Collapsed at Fire
06/27/1948—Sgt. Charles Phillips E-31 Fire-Police Collision
01/11/1949—Sgt. Clifford Bannon E-49 Smoke Inhalation
01/11/1949—FF James Daggert E-40 Smoke Inhalation
05/24/1952—FF Stanley Thornton E-11 Falling Wall
04/11/1954—Capt. Werner Blaess E-1 Heart Attack at Fire
01/26/1955—Lt. Ellsworth Carroll L-14 Heart Attack at Fire
08/23/1959—Capt. Nicholas Konen E-51 Heart Attack at E-51 after Fire
11/24/1959—FF Bruno Koluch E-22 Smoke Inhalation
02/19/1961—FF Stephen Szpunar E-36 Heart Attack at Fire
09/29/1963—FF John Campbell E-34 Heart Attack at Fire
11/10/1963—FF Albert Booth E-56 Car-E-56 Collision
02/27/1965—Capt. Lyle Ingram E-29 Smoke Inhalation
06/29/1965—Sgt. Alex Ori E-48 Heart Attack at Fire
07/28/1967—FF Carl Smith L-11 Gunshot in Riot
08/04/1967—FF John Ashby E-21 Electrocuted

12/20/1969—FF Thomas Killian E-35 Burns
02/22/1970—Capt. Edward Dugelar L-30 Heart Attack at Fire
01/26/1974—FF Terence McHugh E-21 Floor Collapse
03/17/1974—FF Edward Gargol L-26 Fell Off L-26
10/29/1977—TFF Curtis Randolph E-32 Backdraft
12/12/1977—FF Michael Johnson L-11 Tiller Accident
02/21/1981—FF Coleman Tate L-31 Fell Off E-52
03/12/1987—Lt. Paul Schimeck L-10 Fell from Building
03/12/1987—Lt. David Lau E-26 Building Collapse
03/12/1987—TFF Larry McDonald E-26 Building Collapse
06/15/1992—TFF Roland Waters E-60 Building Collapse
02/05/1994—Chief Robert English 5th Battalion Heart Attack at Fire
07/04/1995—FEO John Weingart E-35 Heart Attack at Fire
02/08/2002—FF Steven Olander L-21 Brain Aneurysm
08/14/2005—Sgt. Rodney English Heart Attack
02/07/2007—FEO Joseph Torkos E-17 Accident Responding
11/15/2008—FF Walter Harris E-23 Roof Collapse

In Service to their country:

05/14/1918—Pipeman Joseph Griffin E-8 KIA WWI
06/01/1918—Pipeman Robert R. Lanham E-8 KIA WWI
06/14/1918—Sub. Edward Smalley H-1 KIA WWI
10/27/1918—Pipeman Ambrose B. McMahon H-1 KIA WWI
11/01/1918—Pipeman William Sammon E-17 KIA WWI
08/25/1919—Pipeman Anthony J. Mudloff E-11 KIA WWI
08/23/1943—FF Arthur Peltier E-36 KIA WWII
08/23/1943—FF William Neumann E-8 KIA WWII
08/23/1943—FF Walter Greenberg E-42 KIA WWII
08/23/1943—FF Howard Van Wormer E-49 KIA WWII
05/01/1944—TFF Parke E. Greiner, Jr. E-19 Plane Crash, Navy
03/10/1945—TFF Edward J. Pcyz L-6 KIA WWII
04/29/1945—FF William J. Leonard L-19 KIA WWII
02/15/1951—FF Dallas W. Stephey E-42 KIA Korea
04/18/1961—Lt. William J. Kuenne S-2 Plane Crash, Reserves
07/12/1976—Capt. Thaddeus Potocki E-41 Heart Attack, Reserves

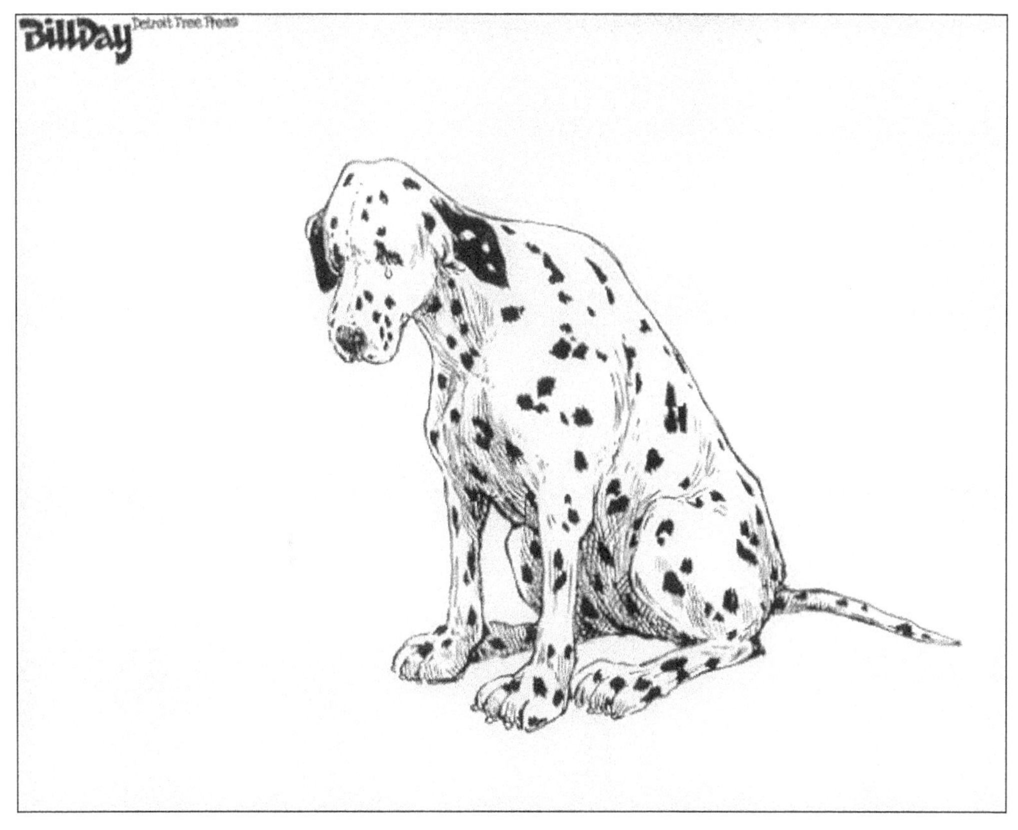

www.ingramcontent.com/pod-product-compliance
Lightning Source LLC
LaVergne TN
LVHW060625110826
845147LV00015B/936

* 9 7 8 1 4 6 7 1 1 5 2 2 3 *